International Conference on Many Particle Spectroscopy of Atoms, Molecules, Clusters and Surfaces 2014

(MPS 2014)

Journal of Physics: Conference Series Volume 601

Metz, France
15 – 18 July 2014

Editor:

L. Ugo Ancarani

ISBN: 978-1-5108-0261-2
ISSN: 1742-6588

Printed from e-media with permission by:

Curran Associates, Inc.
57 Morehouse Lane
Red Hook, NY 12571

Some format issues inherent in the e-media version may also appear in this print version.

Printed by Curran Associates, Inc. (2015)

For permission requests, please contact the Institute of Physics
at the address below.

Institute of Physics
Dirac House, Temple Back
Bristol BS1 6BE UK

Phone: 44 1 17 929 7481
Fax: 44 1 17 920 0979

techtracking@iop.org

Additional copies of this publication are available from:

Curran Associates, Inc.
57 Morehouse Lane
Red Hook, NY 12571 USA
Phone: 845-758-0400
Fax: 845-758-2634
Email: curran@proceedings.com
Web: www.proceedings.com

TABLE OF CONTENTS

Journal of Physics: Conference Series **601** (2015) 011001 doi:10.1088/1742-6596/601/1/011001

International Conference on Many Particle Spectroscopy of Atoms, Molecules, Clusters and Surfaces

15-18 July 2014, Metz, France

Editor

Lorenzo Ugo Ancarani

SRSMC, Université de Lorraine, Metz, France

Preface

This volume contains a collection of contributions from the invited speakers at the 2014 edition of the *International Conference on Many Particle Spectroscopy of Atoms, Molecules, Clusters and Surfaces* held in Metz, France, from 15th to 18th July 2014.

This biennial conference alternates with the ICPEAC satellite *International Symposium on (e,2e), Double Photoionization and Related Topics*, and is concerned with experimental and theoretical studies of radiation interactions with matter. These include many-body and electron-electron correlation effects in excitation, and in single and multiple ionization of atoms, molecules, clusters and surfaces with various projectiles: electrons, photons and ions. More than 80 scientists, from 19 different countries around the world, came together to discuss the most recent progress on these topics. The scientific programme included 28 invited talks and a poster session extending over the three days of the meeting. Amongst the 51 posters, 11 have been selected and were advertised through short talks. Besides, Professor Nora Berrah gave a talk in memory of Professor Uwe Becker who sadly passed away shortly after co-chairing the previous edition of this conference.

Financial support from the Institut Jean Barriol, Laboratoire SRSMC, Groupement de Recherche THEMS (CNRS), Ville de Metz, Metz Métropole, Conseil Général de la Moselle and Région Lorraine is gratefully acknowledged.

Finally, I would like to thank the members of the local committee and the staff of the Université de Lorraine for making the conference run smoothly, the International Advisory Board for building up the scientific programme, the sessions chairpersons, those who gave their valuable time in carefully refereeing the artides of this volume and last, but not least, all participants for contributing to lively and fruitful discussions throughout the meeting.

L.U. Ancarani

February 2015

International Advisory Board

Lorenzo Avaldi (Italy)
Klaus Bartschat (USA)
Jamal Berakdar (Germany)
Nora Berrah (USA)
Igor Bray (Australia)
XiangJun Chen (China)
Claude Dal Cappello (France)
Piero Decleva (Italy)
Mevlut Dogan (Turkey)
Reinhard Dörner (Germany)
Alexander Dorn (Germany)
Omar Fojon (Argentina)
François Frémont (France)
Gustavo Gasaneo (Argentina)
Alexei Grum-Grzhimailo (Russia)
Laszlo Gulyas (Hungary)
Tom Kirchner (Canada)
Konstantin Kouzakov (Russia)
Don Madison (USA)
Fernando Martin (Spain)
Andrew Murray (UK)
Bernard Piraux (Belgium)
Emma Sokell (Ireland)
Elena Magdalena Staicu-Casagrande (France)
Masahiko Takahashi (Japan)
Lokesh Tribedi (India)
Kiyoshi Ueda (Japan)
Jim Williams (Autralia)

Conference photographs

Peer review statement

All papers published in this volume of *Journal of Physics: Conference Series* have been peer reviewed through processes administered by the proceedings Editors. Reviews were conducted by expert referees to the professional and scientific standards expected of a proceedings journal published by IOP Publishing.

High-order harmonic generation driven by plasmonic fields: a new route towards the generation of UV and XUV photons?

M F Ciappina[1], J A Pérez-Hernández[2], L Roso[2], A Zaïr[3] and M Lewenstein[4,5]

[1]Max Planck Institute for Quantum Optics, Hans-Kopfermann-Str. 1, Garching D-85748, Germany
[2]Centro de Láseres Pulsados (CLPU), Parque Científico, 37185 Villamayor, Salamanca, Spain
[3]Blackett Laboratory Laser Consortium, Imperial College London, London SW7 2AZ, United Kingdom
[4]ICFO-Institut de Ciències Fotòniques, Mediterranean Technology Park, 08860 Castelldefels, Spain
[5]ICREA-Institució Catalana de Recerca i Estudis Avançats, Lluis Companys 23, 08010 Barcelona, Spain

E-mail: marcelo.ciappina@mpq.mpg.de

Abstract. We present theoretical investigations of high-order harmonic generation (HHG) resulting from the interaction of noble gases with different kind of temporally and spatially synthesized laser fields. These fields, based on localized surface plasmons, are produced when, for instance, a metal nanoparticle or nanostructure, is illuminated by a few-cycle laser pulse. The enhanced field, which largely depends on the geometrical shape of the metallic nanostructure, has a strong spatial dependency in a scale comparable to the one where the electron dynamics takes place. We demonstrate that the spatial nonhomogeneous character of this laser field plays an important role in the HHG process and leads to a significant increase of the harmonic cutoff energy and modifications in the electron trajectories. The use of metal nanostructures appears to be an alternative way of generating coherent XUV light with a laser field whose characteristics can be spatially synthesized locally.

1. Introduction

There exists nowadays a high demand of coherent light sources in the ultraviolet (UV) to extreme ultraviolet (XUV) spectral range. They configure fundamental tools for basic research, material science, biology, and potentially lithography [1]. One of the main obstacles is a demanding infrastructure for XUV generation and target delivery as well as its low efficiency and duty cycle. Recent experiments based on surface plasmon resonances as light enhancers could potentially solve this problem [2–4].

Field-enhanced high-order-harmonic generation (HHG) using plasmonics fields, generated starting from engineered nanostructures or nanoparticles, requires no extra amplification stages due to the fact that, by exploiting surface plasmon resonances, the incoming electric field can be enhanced by more than 20 dB (corresponding to an increase in the intensity of several orders of magnitude). As a consequence of this enhancement, the threshold laser intensity for HHG generation in noble gases is largely exceeded and the pulse repetition rate remains unaltered. In addition, the high-harmonics radiation generated from each nanosystem acts as a pointlike source, enabling a high collimation or focusing of this coherent radiation by means of (constructive) interference. This fact opens a wide range of possibilities to spatially arrange nanostructures to enhance or shape the spectral and spatial properties of the harmonic radiation in numerous ways [2–4].

Given the nanometric size of the so-called plasmonic 'hot spots', i.e. the spatial region where the electric field reaches its highest intensity, one of the main theoretical assumptions, namely the spatial homogeneity of the driven electric field, should to be softened. Consequently, both the analytical and numerical approaches to study laser-matter processes in atoms and molecules, in particular HHG, need to be modified to treat adequately this new scenario and allow now a spatial dependence in the laser electric field. Several authors have addressed this problem recently [5–14, 20–28, 38]. As we will demonstrate, this new characteristic affects considerably the electron dynamics and its effects are rendered into the observables, in our case the HHG spectra.

In this contribution, we make a sort of brief review about HHG driven by temporally and spatially synthesized laser fields. We first describe succinctly the different theoretical tools, putting emphasis in the modifications needed to tackle this novel scenario. After we present several selected examples, where it is clear that the nohomogeneous character of the laser electric field introduces noticeable modifications both in the HHG spectra and the electron dynamics. Finally we conclude with a short summary.

2. Theory

2.1. Quantum models

Considering the dynamics of an atomic electron in a strong laser field is primarily along the direction of the field, when linearly polarized laser pulses are employed, it is justifiable to model the HHG in a 1D spatial dimension by solving the so-called Time Dependent Schrödinger Equation (1D-TDSE) [7]. In order to model an atom in 1D, it is common to use the so-called soft core potentials, which are of the form $V_{\rm atom}(x) = -1/\sqrt{x^2+a^2}$, where the parameter a allow us to modify the ionization potential of the ground state, fixing it as close as possible to the value of the atom under consideration. The potential due to the laser electric field, linearly polarized along the x axis, is formulated in the dipole approximation and it is modified in order to treat spatially nonhomogeneous fields. As a consequence we write, $V_{\rm laser}(x,t) = E(x,t)x$ where $E(x,t)$ is the laser electric field and now we can include any spatial variation (for details see e.g. [7]). The 1D-TDSE can be solved numerically by using the Crank-Nicolson scheme in order to obtain the time propagated electronic wavefunction $\Psi(x,t)$. Once $\Psi(x,t)$ is found, we can calculate the harmonic spectrum by computing the Fourier transform of the dipole acceleration of the active electron. One of the main advantages of the 1D-TDSE is that we are able to include any functional form for the spatial variation of the plasmonic field. For instance, we have implemented linear [7] and real (parabolic) plasmonic fields [7], as well as near-fields with exponential decay [25]. For all the cases we have observed noticeable differences in the HHG spectra as we will show in Section 3.

A natural extension of the above described approach is to solve the TDSE in its full dimensionality and to include in the laser-electron potential the spatial variation of the laser electric field. For only one active electron we need to deal with 3 spatial dimensions and, due to the cylindrical symmetry of the problem, we are able to separate the electronic wavefunction in spherical harmonics, Y_l^m, and take into account only terms with $m = 0$. Typically several hundreds of angular momenta l should to be considered and we could recognize the time evolution of each of them as a 1D problem. We use a Crank-Nicolson method implemented on a splitting of the time-evolution operator that preserves the norm of the wave function for the time propagation, similar to the 1D-TDSE case. We base our studies in helium because a majority of experiments in HHG are carried out in noble gases, but other atoms could be easily implemented. After time propagation of the electronic wavefunction, the HHG spectra can be computed in an analogous way as in the case of the 1D-TDSE. Due to the complexity of the problem, we were able to carry out studies only with nonhomogeneous fields with linear spatial variations in the 3D-TDSE, but this example is enough to confirm how a spatial nonhomogeneous field strongly modifies the HHG spectra (for more details see [14]).

2.2. Semiclassical approach

An alternative approach to compute high-harmonic spectra for atoms in intense laser pulses is the Strong Field Approximation (SFA) or Lewenstein model [39]. The main ingredient of this approach is the evaluation of the time-dependent dipole moment $\mathbf{d}(t)$. Within the single active electron (SAE) approximation it can be calculated starting from the ionization and recombination transition matrices combined with the classical action of the laser-ionized electron moving in the laser field. The SFA approximation has a direct interpretation in terms of the so-called three-step or simple man's model [39]. The first step is the strong-field ionization of the atom or molecule as a consequence of the nonperturbative interaction with the coherent electromagnetic radiation. Next, the classical movement of the electron in the laser electric field defines the second step of the model. Finally, the third step in the sequence arises when the laser-ionized electron is steered back to its origin, due to the inversion in the direction of the laser electric field, recombining with the parent ion under the emission of a high-energy photon. One of the main features of the HHG process is the coherence of the emitted radiation, which, amongst other uses, opens the possibility of generating attosecond pulses [1]. Implicitly the Lewenstein model deals with spatially homogeneous electric and vector potential fields, i.e. fields that do not experience variations in the region where the electron dynamics takes place. In order to consider spatially nonhomogeneous fields, the SFA approach needs to be modified accordingly, i.e. the ionization and recombination transition matrices, joint with the classical action, should to take into account this new feature of the laser field (for further details see [7,8]).

2.3. Classical framework

Relevant information, for instance the HHG cutoff and properties of the electron trajectories, can be obtained solving the classical Newton-Lorentz equation for an electron moving in a linearly polarized electric oscillating field, i.e. using fully classical considerations. The following conditions are commonly set (the resulting model is also known as the simple man's model): i) the electron starts with zero velocity at the origin at time $t = t_i$, defining the ionization time t_i, i.e., $x(t_i) = 0$ and $\dot{x}(t_i) = 0$; (ii) when the laser electric field reverses its direction, the electron returns to its initial position, i.e., recollides with the parent ion, at a later time, $t = t_r$ (the so-called recollision time), i.e. $x(t_r) = 0$. The kinetic energy of the laser-ionized electron at the recollision (return) time t_r can be obtained from the usual formula $E_k(t_r) = \dot{x}(t_r)^2/2$, and, finding the value of t_r (as a function of t_i) that maximizes this energy, we find that the HHG cutoff holds $n_c\omega = 3.17U_p + I_p$, where n_c is the harmonic order at the cutoff, ω is the laser frequency, U_p is the ponderomotive energy ($U_p = I/4\omega^2$, with I being the laser intensity in a.u.), and I_p is the ionization potential of the atom or molecule under consideration. It is worth of mention that the HHG cutoff will be extended when spatially inhomogeneous fields are employed as we will see in Section 3. We have solved numerically the Newton equation for an electron moving in a linearly polarized (in the x-axis) electric field with the same parameters used in the quantum models in order to compare the outcomes of both (see Section 3). Specifically, we find the numerical solution of $\ddot{x}(t) = -\nabla_x V_{\text{laser}}(x,t)$ where $V_{\text{laser}}(x,t)$ was defined in Section 2.1. For fixed values of ionization times t_i, it is possible to obtain the classical trajectories and to numerically calculate the times t_r for which the electron recollides with the parent ion. In

Journal of Physics: Conference Series **601** (2015) 012001 doi:10.1088/1742-6596/601/1/012001

addition, once the ionization time t_i is fixed, the full electron trajectory is completely determined (for more details about the classical model see [28]).

3. Selected results

In the following sub-sections we present a brief summary of the results reported in several articles published recently. In these works, different noble gases (He, Ar and Xe) are used as atomic targets located in the vicinity of metal nanotips and nanoparticles and the HHG generated by them was studied and characterized.

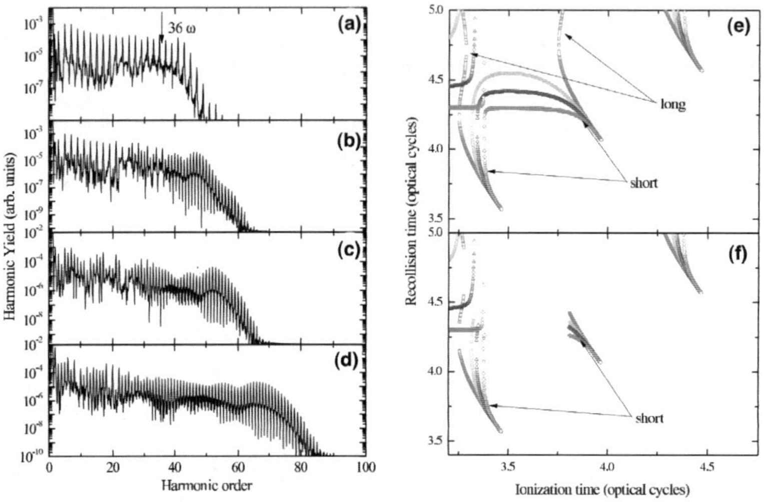

Figure 1. HHG spectra for a model atom with a ground-state energy, $\mathcal{E}_{\rm GS} = -0.67$ a.u. generated using the 1D-TDSE approach. The laser parameters are $I = 2 \times 10^{14}$ W·cm^{-2} and $\lambda = 800$ nm. We have used a trapezoidal shaped pulse with two optical cycles turn on, i.e. $n_{\rm on} = 2$, and turn off, i.e. $n_{\rm off} = 2$, and a plateau with six optical cycles, i.e. $n_{\rm p} = 6$ (10 optical cycles in total, i.e. approximately 27 fs). The arrow indicates the cutoff predicted by the semiclassical model [39]. Panel (a): homogeneous case, (b): $\varepsilon = 0.01$ (100 a.u), (c): $\varepsilon = 0.02$ (50 a.u) and (d): $\varepsilon = 0.05$ (20 a.u). The parameter ε controls the *strength* of the inhomogeneity and for the linear case has dimension of inverse length. The values between brackets correspond to the inhomogeneity regions (see [7] for more details). In panels (e) and (f) is shown the dependence of the semiclassical trajectories on the ionization and recollision times for different values of ε and for the non confined case, panel (e) and the confined case, panel (f), respectively. Red squares: homogeneous case $\varepsilon = 0$; green circles: $\varepsilon = 0.01$; blue triangles: $\varepsilon = 0.02$ and magenta diamonds: $\varepsilon = 0.05$. This figure has been adapted from [7].

3.1. HHG driven by spatially (linear) nonhomogeneous fields and electron confinement

In this sub-section we summarize the study carried out in [7] where it is shown that both the inhomogeneity of the local fields and the confinement of the electron movement, play an important role in the HHG process and lead to the generation of even harmonics and a significant increase in the HHG cutoff, more pronounced for the longer wavelengths cases studied. In order to understand and characterize these new HHG features we employ two of the different approaches mentioned above: the numerical solution of the 1D-TDSE (see panels (a)-(d) in

Fig. 1) and the semiclassical approach known as Strong Field Approximation (SFA). Both approaches predict comparable results and reproduce satisfactorily the new features, but by employing the semiclassical arguments (see panels (e) and (f) in Fig. 1) behind the SFA and time-frequency analysis tools (Fig. 2), we are able to fully understand the reasons of the cutoff extension. A more detailed study can be found in [7].

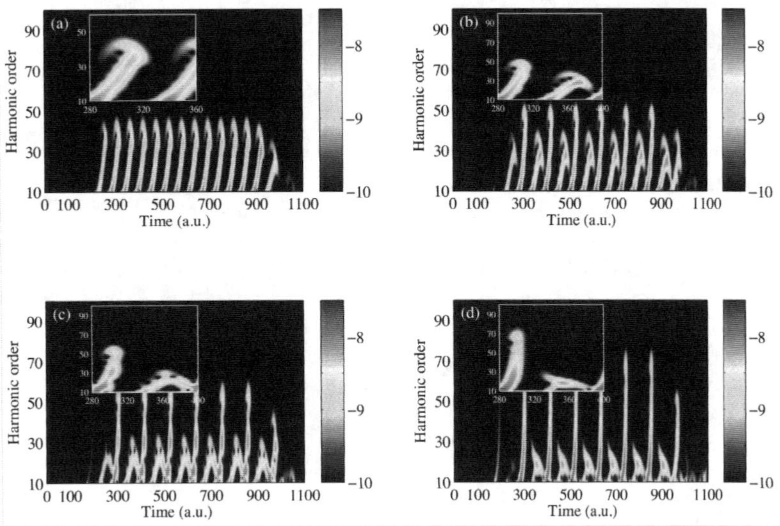

Figure 2. Panels (a)-(d): Gabor analysis for the corresponding HHG spectra of panels (a)-(d) of Fig. 1. The zoomed regions in all panels show a time interval during the laser pulse for which the complete electron trajectory, from birth time to recollision time, falls within the pulse plateau. In panels (a)-(d) the color scale is logarithmic. This figure has been adapted from [7].

3.2. HHG driven by real nonhomogeneous fields

In this sub-section we include numerical simulations of HHG in Ar produced by the fields generated when a gold bow-tie nanostructure is illuminated by a short laser pulse of long wavelength (see [9] for more details). The functional form of these fields is extracted from finite element simulations using both the complete geometry of the metal nanostructure and the laser pulse characteristics (see Fig. 3(a)). We use the numerical solution of the TDSE in reduced dimensions to predict the HHG spectra. We observe an extension in the harmonic cutoff position that could lead to the production of XUV coherent laser sources and open the avenue to the generation of attosecond pulses. It is shown in Fig. 3(c) that this new feature is a consequence of the combination of a spatial nonhomogeneous electric field, which modifies substantially the electron trajectories, and the confinement of the electron dynamics. Furthermore, our numerical results are supported by time-analysis and classical simulations. A more pronounced increment in the harmonic cutoff, in addition with an appreciable growth in the conversion efficiency, could be attained optimizing the nanostructure geometry and by choosing the adequate materials. These degrees of freedom could pave the way to tailor the harmonic spectra according to specific requirements.

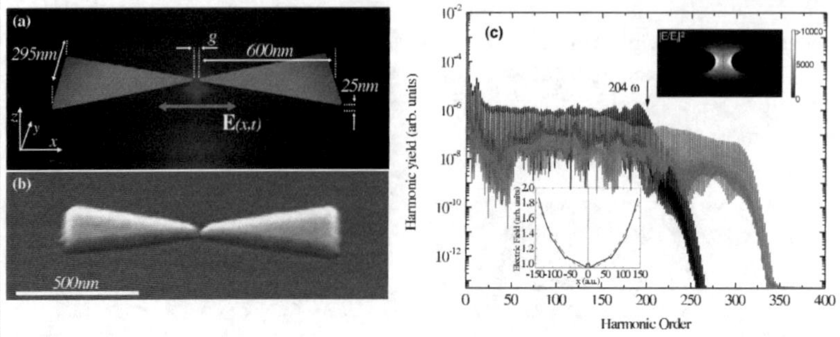

Figure 3. (a) Schematic representation of the geometry of the considered nanostructure. A gold bow-tie antenna resides on a glass substrate (refractive index $n = 1.52$) with superstate medium of air ($n = 1$). The characteristic dimensions of the system and the coordinate system used in the 1D-TDSE simulations are shown. (b) SEM image of a real gold bow-tie antenna. (c) High-order harmonic generation (HHG) spectra for Ar (Ionization potential $\mathcal{E}_{\rm GS} = -0.58$ a.u.), driven by a laser pulse with wavelength $\lambda = 1800$ nm and intensity $I = 1.25 \times 10^{14}$ W·cm^{-2} at the center of the gap $x = 0$. We have used a trapezoidal shaped pulse with three optical cycles turn on, i.e. $n_{\rm on} = 3$, and turn off, i.e. $n_{\rm off} = 3$, and a plateau with four optical cycles, i.e. $n_{\rm p} = 4$ (about 60 fs). The gold bow-tie nanostructure has a gap $g = 15$ nm (283 a.u.). The black line indicates the homogeneous case while the red line indicates the nonhomogeneous case. The arrow indicates the cutoff predicted by the semiclassical model for the homogeneous case [39]. The top left inset shows the functional form of the electric field $E(x,t)$, where the solid lines are the raw data obtained from the finite element simulations and the dashed line is a nonlinear fitting. The top right inset shows the intensity enhancement in the gap region of the gold bow-tie nanostructure. This figure has been adapted from [9].

3.3. HHG driven by temporal and spatial synthesized fields

In this sub-section we present a brief report of the results published in [14]. In short, numerical simulations of HHG in He atoms using a temporal and spatially synthesized laser field are considered using the full 3D-TDSE. The field resulting from both time and space synthesis, provides a new route for the generation of photons at energies beyond the carbon K-edge using $\lambda = 800$ nm laser pulses, which can be obtained from conventional Ti:Sapphire laser sources, now relatively common amongst the strong field physics community. The temporal synthesis is performed using two few-cycle laser pulses delayed in time [15]. Few-cycle laser pulses are now routinely used for attoscience applications and are produced using a post-compression stage [16, 17]. The spatial synthesis is obtained by using a nonhomogeneous laser field [5–7] produced when a laser beam is focused in the vicinity of a metal nanostructure or nanoparticle. In particular, the temporal synthesized laser field is described by $E_1(t) = E_0 \sin^2\left(\frac{\omega t}{2N}\right)\sin(\omega t)$ and $E_2(t,\tau) = E_0 \sin^2\left(\frac{\omega(t-\tau)}{2N}\right)\sin(\omega(t-\tau)+\phi)$ where E_0 is the laser electric field amplitude in atomic units ($E_0 = \sqrt{I/I_0}$ with $I_0 = 3.5 \times 10^{16}$ W/cm^2), $\omega = 0.057$ a.u. the laser frequency corresponding to $\lambda = 800$ nm, T the laser period, N the total number of cycles in the pulse, ϕ the carrier-envelope phase (CEP) and τ the time delay between the pulses. We consider the case $N = 4$ and $\phi = 0$ and $\tau = 1.29T$ (see [15] for more details). It is worth to mention, however, that there exists another complementary ways to perform the temporal synthesis, for instance, using chirped input laser pulses [18, 19].

Regarding the spatial synthesis, the nonhomogeneous spatial distribution of the laser field can be obtained experimentally by using the laser field as produced by nano-plasmonic antennas [2, 5–7], metallic waveguides [3], metal [29, 30] and dielectric nanoparticles [31] or metal nanotips [32–36]. The main feature of these fields is that they are no longer spatially homogeneous in the region where the electron dynamics takes place. Very recently, studies about how HHG spectra are modified due to spatial nonhomogeneous fields have been published [5–7,9] leading to the so-called 'HHG driven by resonant plasmon field enhancement' showing the growing interest of such new subject.

The coupling between the atom and the laser pulse, linearly polarized along the z axis, is modified in order to treat spatially nonhomogeneous fields as: $V_{\text{laser}}(z,t,\tau) = \tilde{E}(z,t,\tau)\,z$, with $\tilde{E}(z,t,\tau) = E(t,\tau)(1+\beta z)$, where $V_{\text{laser}}(z,t,\tau)$ represents the laser-atom interaction in the length gauge and $E(t,\tau) = E_1(t) + E_2(t,\tau)$ is the temporal synthesized laser field. The parameter β defines the strength of the nonhomogeneity and the dipole approximation is preserved because $\beta \ll 1$.

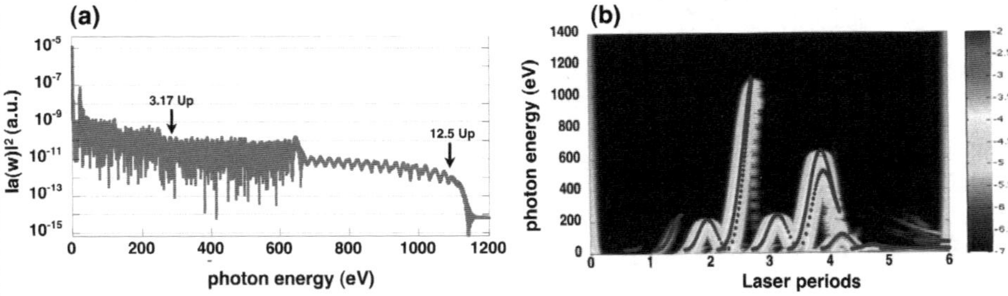

Figure 4. (a) 3D-TDSE harmonic spectrum for a He atom driven by the spatially and temporally synthesized pulse described above with $\beta = 0.002$. The plasmonic enhanced intensity is $I = 1.4 \times 10^{15}$ W/cm^2. (b) Time-frequency analysis obtained from the 3D-TDSE and superimposed (in brown) with the classical rescattering energies for the same parameters used in (a). This figure has been adapted from [14].

The linear functional form for the spatial nonhomogeneity described above could be obtained by engineering adequately the geometry of plasmonic nanostructures and adjusting the laser parameters in such a way that the laser-ionized electron feels only a linear spatial variation of the laser electric field when in the continuum (see e.g. [9] and references therein). The harmonic spectrum then obtained in He for $\beta = 0.002$ is presented in Fig. 4(a). We can observe a considerable cut-off extension up to $12.5U_p$ which is much larger when compared with the double pulse configuration employed alone (it leads only to a maximum of $4.5U_p$ [15]). This large extension of the cut-off is therefore a signature of the combined effect of the double pulse and the spatially nonhomogeneous character of the laser electric field. For this particular value of the laser peak intensity (1.4×10^{15} W/cm^2) the highest photon energy is greater than 1 keV. Note that the quoted intensity is actually the plasmonic enhanced intensity, not the input laser intensity. The latter could be several orders of magnitude smaller, according to the plasmonic enhancement factor (see e.g. [2,3]) and will allow the nanoplasmonic target to survive to the interaction with the incoming laser pulse. In order to confirm the underlying physics highlighted by the classical trajectories analysis, we retrieved the time-frequency distribution of the calculated dipole (from the 3D-TDSE) corresponding to the case of the spectra presented in Fig. 4(a) using a wavelet analysis. The result is presented in Fig. 4(b) where we have superimposed the calculated classical rescattering energies (in brown) to show the excellent agreement between the two theoretical approaches. The consistency of the classical calculations with the full quantum approach is clear and confirms the mechanism of the generation of this $12.5U_p$ cut-off extension. In addition, the HHG spectra exhibit a clean supercontinuum as a result of the trajectory selection on the recollision time, consequences of employing the combination of a temporally and spatially synthesized laser field.

3.4. HHG driven by plasmonic near-fields

This sub-section includes a brief summary of the results reported in [20]. In such contribution it is shown how the HHG spectra from Xe atoms are modified by using a plasmonic enhanced near-field generated when a metal nanoparticle is illuminated by a short laser pulse. We use the function given by [30] to define the spatial nonhomogeneous laser electric field $E(x,t)$, i.e. $E(x,t) = E_0\, f(t)\, \exp(-x/\chi)\sin(\omega t + \phi)$ where E_0, ω, $f(t)$ and ϕ are the peak amplitude, the laser field frequency, the field envelope and the phase, respectively. The functional form of the resulting laser electric field is extracted from attosecond streaking experiments and incorporated both in our quantum and classical mechanical approaches. In this specific case the spatial dependence of the plasmonic near-field is given by $\exp(-x/\chi)$ and it depends on both the size and the material of the spherical nanoparticle used. $E(x,t)$ is valid for x outside of the metal nanoparticle, i.e. $x \geq R_0$, where R_0 is its radius and it is important to note that the electron motion stands in the region $x \geq R_0$ with $(x + R_0) \gg 0$. We consider the laser field having a $\sin^2$ envelope: $f(t) = \sin^2\left(\frac{\omega t}{2n_{\rm p}}\right)$, where $n_{\rm p}$ is the total number of optical cycles, i.e. the total pulse duration is $\tau = 2\pi n_{\rm p}/\omega$. The harmonic yield of the atom is obtained by Fourier transforming the dipole acceleration $a(t)$ of the electronic wavepacket.

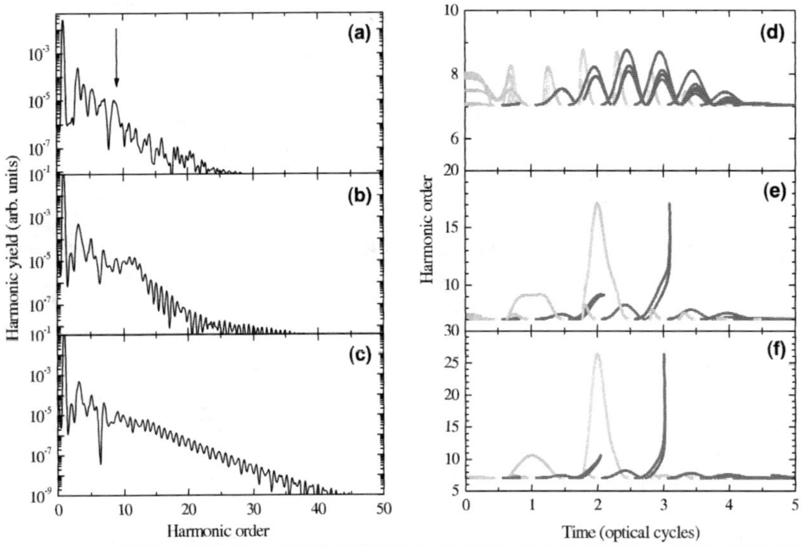

Figure 5. HHG spectra for Xe, laser wavelength $\lambda = 720$ nm and intensity $I = 2 \times 10^{13}$ W·cm^{-2}. We use a $\sin^2$ shaped pulse with $n_{\rm p} = 5$. Panel (a) shows the homogeneous case, panel (b) $\chi = 50$ and panel (c) $\chi = 40$. The arrow in panel (a) indicates the cut-off predicted by the semiclassical model [39]. Panels (d)-(f) represent the corresponding total energy of the electron (expressed in harmonic order) driven by the laser field calculated from Newton's second law and plotted as a function of the t_i (green (light gray) circles) or the t_r (red (dark gray) circles). This figure has been adapted from [20].

Figure 5 panels (a)-(c) depict the harmonic spectra for Xenon generated by a laser pulse with $I = 2 \times 10^{13}$ W/cm^2, $\lambda = 720$ nm and a $\tau = 13$ fs, i.e $n_{\rm p} = 5$ (this pulse corresponds to an intensity envelope of ≈ 4.7 fs FWHM) [30]. For the case of a homogeneous field no harmonics beyond the 9$^{\rm th}$ order are observed. The spatial decay constant χ quantifies the nonhomogeneity due to the nanoparticle and it varies together with its size and the material (metal) employed. Changing χ is therefore equivalent to choosing the type of nanoparticle used, which leads to overcome the semiclassically predicted cutoff limit and reach higher harmonic orders. For instance with $\chi = 40$ and $\chi = 50$ harmonics in the mid 20s (panel c) and well above the 9$^{\rm th}$ (a clear cut-off at $n_c \approx 15$ is visible) (panel b), respectively, are observed. A change in the harmonic periodicity, related to the breaking of symmetry imposed by the induced non-homogeneity, is clearly visible as well.

We now consider the semiclassical simple man's (SM) model [37,39] in order to characterize the harmonic cut-off extension. This new feature may appear due to the combination of several factors (for details see [7,9]). It is well established that $n_c = (3.17U_p + I_p)/\omega$, where n_c is the harmonic order at the cut-off and U_p the ponderomotive energy. We solve numerically the Newton-Lorentz equation for an electron moving in an oscillating electric field with the same parameters used in the TDSE-1D calculations, i.e. $\ddot{x}(t) = -\nabla_x V_{\rm laser}(x,t) = -E(x,t)(1 - \frac{x(t)}{\chi})$, and consider the SM model initial conditions: the electron starts at position zero at $t = t_i$ (the ionization time) with zero velocity, i.e. $x(t_i) = 0$ and $\dot{x}(t_i) = 0$. When the laser electric field reverses its direction, the electron returns to its initial position (i.e. the electron *recollides* or recombines with the parent ion) at a later time $t = t_r$ (the recollision time), i.e. $x(t_r) = 0$. The electron kinetic energy at the t_r is calculated from: $E_k(t_r) = \frac{\dot{x}(t_r)^2}{2}$ and finding the t_r (as a function of t_i) that maximizes E_k, n_c is also maximized.

In panels (d)-(f) of Fig. 5 we show the dependence of the harmonic order upon the t_i and t_r, calculated from $n = (E_k(t_{i,r}) + I_p)/\omega$ as for the cases (a)-(c) of Fig. 5, respectively. Panels (a), (b) and (c) depict the cases of ($\chi \to \infty$) (homogeneous field), $\chi = 50$ and $\chi = 40$, respectively. Panels (b) and (c) show how the nonhomogeneity character of the laser field modifies considerably the electron trajectories towards an extension of the n_c. Limits clearly present at $n_c \sim 18\omega$ (28 eV) and $n_c \sim 27\omega$ (42 eV) for $\chi = 50$ and $\chi = 40$, respectively, are observed. These last two cut-off values are indeed consistent with the quantum mechanical calculations presented in panels (b) and (c) of Fig. 5.

Both models, classical and quantum, report an extension in the n_c position that could lead to the production of XUV coherent laser sources and opening the avenue to the generation of attosecond pulses. This new feature is a consequence of the induced laser field nonhomogeneity only, which modifies substantially the electron trajectories. A more pronounced increment in the harmonic cut-off, in addition with an appreciable growth in the conversion efficiency, could be reached varying, for instance, both the radius and the material (metal) of the spherical nanoparticles. These new degrees of freedom could pave the way to enhance the harmonic spectra reaching the XUV regime with modest input laser intensities.

4. Conclusions

We studied high-order-harmonic generation (HHG) driven by temporal and spatial synthesized fields. These fields are not merely a theoretical speculation but are present in a vicinity of a metal nanostructure or nanoparticle, of nanometric dimensions, when it is illuminated by a short and intense laser pulse. We have shown several prominent examples where it is possible to infer that the spatial nonhomogeneous character of the laser electric field introduces a new degree of freedom, which, manipulated adequately, could open the possibility to generate efficiently coherent radiation in the UV and XUV range. Regarding the experimental feasibility, important challenges still remains about the possibility to achieve high energetic photons via spatial nonhomogeneous fields and, in some aspects this is not free of controversy [40]. However, we would like to stress out that the theory envisions that this process is physically feasible. Our predictions are based on the convergence of three different approaches, quantum, semiclassical and classical, and consequently the robustness of our results appears to be settled.

Acknowledgments

We acknowledge the financial support of the ERC Advanced Grants QUAGATUA and OSYRIS, Alexander von Humboldt Foundation (M L); J A P-H. and L R acknowledge support from Laserlab-Europe (Grant No. EU FP7 284464) and the Spanish Ministerio de Economía y Competitividad (FURIAM Project FIS2013-47741-R). This research has been partially supported by Fundació Privada Cellex. M F C acknowledges Ferenc Krausz for useful comments and suggestions.

References

[1] Krausz F and Ivanov M 2009 *Rev. Mod. Phys.* **81** 163
[2] Kim S et al 2008 *Nature (London)* **453** 757
[3] Park I-Y et al 2011 *Nat. Phot.* **5** 677
[4] Pfullmann N et al 2013 *New J. Phys.* **15** 093027
[5] Husakou A et al 2011 *Phys. Rev. A* **83** 043839
[6] Yavuz I et al 2012 *Phys. Rev. A* **85** 013416
[7] Ciappina M F et al 2012 *Phys. Rev. A* **85** 033828
[8] Shaaran T et al 2012 *Phys. Rev. A* **86** 023408
[9] Ciappina M F et al 2012 *Opt. Express* **20** 26261
[10] Ciappina M F et al 2012 *Phys. Rev. A* **86** 023413
[11] Shaaran T et al 2012 *J. Mod. Opt.* **59** 1634
[12] Shaaran T et al 2013 *Ann. Phys. (Berlin)* **525** 97
[13] Fetić B et al 2013 *Ann. Phys. (Berlin)* **525** 107
[14] Pérez-Hernández J A et al 2013 *Phys. Rev. Lett.* **110** 053001
[15] Pérez-Hernández J A et al 2009 *J. Phys. B* **42**, 134004
[16] Sansone G et al 2006 *Science* **314**, 443
[17] Eckle P et al 2008 *Nat. Phys.* **4** 565-570
[18] Carrera J J et al 2006 *Phys. Rev. A* **74** 023404
[19] Yang Y-Y et al 2012 *Opt. Express* **21** 2195
[20] Shaaran T et al 2013 *Phys. Rev. A* **87** 041402
[21] Shaaran T et al 2013 *Phys. Rev. A* **87** 053415
[22] Yavuz I 2013 *Phys. Rev. A* **87** 053815
[23] Ciappina M F et al 2013 *Phys. Rev. A* **87** 063833
[24] Luo J et al 2013 *J. Phys. B* **46** 145602
[25] Ciappina M F et al 2013 *Laser Phys. Lett.* **10** 105302
[26] Ciappina M F et al 2014 *Phys. Rev. A* **89** 013409
[27] Feng L Q et al 2013 *Phys. Plasmas* **20** 122307
[28] Ciappina M F et al 2014 *Comp. Phys. Comm.* **185** 398
[29] Zherebtsov S et al 2011 *Nat. Phys.* **7** 656
[30] Süßmann F and Kling M F 2011 *Proc. of SPIE* **8096** 80961C
[31] Süßmann F and Kling M F 2011 *Phys. Rev B* **84** 121406(R)
[32] Hommelhoff P et al 2006 *Phys. Rev. Lett.* **96** 077401
[33] Schenk M et al 2010 *Phys. Rev. Lett.* **105** 257601
[34] Krueger M et al 2011 *Nature (London)* **475** 78
[35] Krueger M et al 2012 *J. Phys. B* **45**, 074006
[36] Herink G et al 2012 *Nature (London)* **483**
[37] Corkum P B 1993 *Phys. Rev. Lett.* **71** 1994
[38] Husakou A and Herrmann J 2014 *Phys. Rev. A* **90** 023831
[39] Lewenstein M et al 1994 *Phys. Rev. A* **49** 2117
[40] Sivis M et al 2012 *Nature (London)* **485** E1-E3

Journal of Physics: Conference Series **601** (2015) 012002 doi:10.1088/1742-6596/601/1/012002

Observing interference effect in binary (e, 2e) of molecules

Xiangjun Chen[1, 2, *], Jing Yang[1, 2], Xu Shan[1, 2], Zhe Zhang[1], Qiguo Tian[1] and Enliang Wang[1, 2]

[1]Hefei National Laboratory for Physical Sciences at Microscale and Department of Modern Physics, University of Science and Technology of China, Hefei, Anhui, 230026, China
[2]Synergetic Innovation Center of Quantum Information and Quantum Physics, University of Science and Technology of China, Hefei, Anhui 230026, China

E-mail: xjun@ustc.edu.cn

Abstract. We present investigations on the two-center and multi-center interference effects of molecules in binary (e, 2e) experiments. The high energy resolution electron momentum spectroscopy (EMS) measurements on H_2 are reported with final vibrational states resolved. The experimental momentum profiles for ionization transitions to the individual final vibrational states of the ion are obtained. The measured and calculated vibrational ratios of the cross sections deviate from Franck-Condon principle, which can be ascribed to the Young-type two-center interference. Furthermore, with the help of our latest version of EMS spectrometer which has considerably higher sensitivity and much wider momentum range from 0 to 8 a.u., we are able to extend our observations to multi-center interference effect in high symmetry molecules like NF_3 and CF_4 with several oscillation periods included.

1. Introduction

As we all know, particle-wave duality of matter particles plays key role in quantum mechanics. It is one of the prominent conceptual deviations from the classical physics. This revolutionary concept has been directly demonstrated by the beautiful electron double-slit experiment by Clause Jönsson in 1961 [1]. Since then, double-slit experiments have shown the wave character of increasingly larger quantum objects, including fullerenes (buckyballs) [2] and huge organic molecules [3]. Another way to realize the double-slit experiment is the coherent superposition of electrons emitted from two indistinguishable atoms in diatomic molecules [4-22]. This is often referred to as molecular double-slit interference, which was first suggested by Cohen and Fano [4] in 1966 in photoionization of diatomic molecules N_2 and O_2. Such interference effects will lead to the energy- or angle-dependent oscillations in cross sections. It took 35 years before this two-center interference effect was unambiguously proven for H_2 in the ionizations by heavy ions [5], and another 4 more years in 2005 to find evidence in photoionization of N_2 [10].

The interference effect in dipole (e, 2e) ionization of H_2 was first predicted by theoretical calculations in 2003 [17] and then observed by experiments employing coplanar asymmetric kinematics at intermediate energies [18, 19]. The effect was revealed from the suppression or enhancement of the forward (binary) or backward (recoil) scattering peaks as compared to helium at same kinematics. In

* To whom any correspondence should be addressed.

fact, it would be more straightforward to observe Young-type interference in binary (e, 2e). For binary (e, 2e), we have (1) high electron impact energy; (2) larger momentum transfer to ejected electron. If the Bethe ridge condition is satisfied that the incident electron transfers all the lost momentum to target electron, the triple differential cross section (TDCS) of (e, 2e) [23] is then directly linked to the square modulus of the single-electron wave function in momentum space and can be used to directly 'imaging' the electron momentum distributions for individual orbitals. The relevant technique is usually terminized as electron momentum spectroscopy (EMS).

For the simplest diatomic molecule H_2, as a simple approximation, the molecular orbital (MO) ψ can be expressed by a linear combination of atomic orbitals (LCAO):

$$\psi = \frac{1}{\sqrt{2}}(1s_A + 1s_B)\,, \tag{1}$$

where $1s_A$ and $1s_B$ are two identical $1s$ atomic orbitals centered on atoms A and B of the molecule. It is easy to obtain the TDCS [17]

$$\sigma_{EMS}^{(3)} \cong 2[1+\sin(qR_0)/(qR_0)]\sigma_H^{(3)}\,, \tag{2}$$

where $\sigma_H^{(3)}$ is the atomic TDCS, R_0 is the internuclear distance at equilibrium and q is the magnitude of the recoil momentum of residual ion which is equal to p, the magnitude of the momentum of orbital electron, under the EMS condition. In principle, we can directly observe the interference factor $1+\sin(pR_0)/(pR_0)$ by plotting the ratio of TDCSs of molecule and atom as a function of momentum p. Such phenomena is also called bond oscillation [24, 25] and its prediction can be traced back to the early 1940s [26]. However, for H_2, it is difficult to observe such interference effect for two reasons. Firstly, the momentum range for a conventional EMS instrument is limited to 0 ~ 3.0 a.u. A much wider range of momenta up to $2\pi/R_0$ = 4.5 a.u.is at least required to observe a complete period of bond oscillation in H_2, having the internuclear distance R_0 of 1.4 a.u. Secondly, the EMS cross section decreases very rapidly as momentum increases. In order to have clear observations of interference effect, one way is to choose a molecule having larger internuclear distance. Most recently, the multi-centre interference effect was observed in binary (e, 2e) experiment for the three outermost MOs of CF_4 [22], each of which are consisted of a combination of non-bonding $2p$ AOs located on the four F atoms. The F-F internuclear distance is 4.02 a.u. and the period $2\pi/R_0$ = 1.6 a.u. is well located in the momentum range of EMS. Another way to observe the interference effect is to compare the interference factor at different internuclear distance. In this talk, we present our first measurement [27] on vibrationally resolved EMS of H_2 by a high-resolution (e, 2e) spectrometer [28]. The experimental momentum profiles for ionization transitions to the individual final vibrational states of the ion are obtained. By choosing different vibrational states, we equivalently change the internuclear distance. The measured and calculated vibrational ratios of cross sections reveal obvious deviations from Franck-Condon values. Such deviations can be ascribed to the Young-type two-center interference. Furthermore, with the help of our latest version of EMS spectrometers [29] which have considerably higher sensitivity and much wider momentum range from 0 to 8 a.u., we are able to extend our observations to multi-center interference effect in high symmetry molecules like NF_3 and CF_4.

2. Background

EMS is based on the *(e, 2e)* experiment in which an electron from target atom or molecule is cleanly knocked out by a high-energy incident electron and the residual ion acts as a spectator. From energy and momentum conservation, the binding energy ε_f and the momentum $\boldsymbol{p}$ of the target electron are given by

$$\varepsilon_f = E_0 - E_a - E_b \tag{3}$$

$$\boldsymbol{p} = \boldsymbol{p}_a + \boldsymbol{p}_b - \boldsymbol{p}_0 \tag{4}$$

where E_i, $\boldsymbol{p}_i$ (i = 0, a, b) are kinetic energies and momenta of the incident and two outgoing electrons, respectively.

Within the binary-encounter approximation, as well as the plane wave impulse approximation (PWIA), the TDCS for (e, 2e) ionization is [23]

$$\frac{d^3\sigma}{d\Omega_a d\Omega_b dE_b} = (2\pi)^4 \frac{p_a p_b}{p_0} f_{ee} \sum_{av} |\langle \boldsymbol{p} I | G \rangle|^2 \quad (5)$$

Here f_{ee} is the electron-electron collision factor which is essentially constant in EMS conditions. Therefore the cross section is proportional to a structure term which is the square of overlap between the initial target state G and final ion state I. $\sum_{av}$ denotes a sum for final states and average for initial states that are not resolved in the experiment and are considered as degenerate. For molecular target the initial state G and final state I can be described in terms of Born-Oppenheimer approximation which is a product of separate electronic, vibrational, and rotational functions [23]:

$$|G\rangle = |0 V_\mu D_\nu\rangle \quad (6)$$

$$|I\rangle = |i V'_{\mu'} D'_{\nu'}\rangle \quad (7)$$

where V_μ and D_ν are the vibrational and rotational functions for the initial state. The indices μ and ν represent quantum numbers that specify the vibrational and rotational states respectively. Final vibrational and rotational quantities are denoted by primes. The notations 0 and i represent the electronic states of target and ion. At room temperature the target is in its vibrational ground state V_0 ($\mu = 0$).

For the conventional EMS spectrometer, the energy resolution (typically 1~2 eV) cannot resolve the final vibrational and rotational states. The cross section reduces to

$$\sum_{av} |\langle \boldsymbol{p} I | G \rangle|^2 = \int \frac{d\Omega}{4\pi} \langle V_0 || \langle \boldsymbol{p} i | 0 \rangle|^2 | V_0 \rangle \quad (8)$$

For the H_2 molecule Dey *et al.*[30] showed that the vibrational average gives the same results as taking the electronic functions at their equilibrium nuclear geometry. This approximation has subsequently been justified by agreement with a wide range of experiments. So the cross section further reduces to

$$\sigma^{(3)}_{EMS} = \frac{d^3\sigma}{d\Omega_a d\Omega_b dE_b} \propto \int \frac{d\Omega}{4\pi} |\langle \boldsymbol{p} i | 0 \rangle|^2 \quad (9)$$

The integral in equation (9) is known as the spherically averaged electron momentum distribution, or electron momentum profile.

3. Two-center interference effects in vibrationally resolved (e, 2e) of H_2

Figure 1(a) illustrates the ionization process involved in this work - vibronic transition from $X^1\Sigma_g^+$ ground state of H_2 to $^2\Sigma_g^+(1s\sigma_g)$ ground state of H_2^+ ion. At room temperature the ionization starts from $\mu = 0$ vibrational ground state of target to a certain μ' vibrational state of ion. Figure 1(b) shows the ionization energy spectrum obtained by high-resolution photoelectron spectroscopy (PES) [31]. The final vibrational states of H_2^+ are well resolved and the relative intensities of the individual vibrational peaks are proportional to the Franck-Condon factors. The binding energy spectrum (BES) measured in this work is shown in figure1(c). Although the energy resolution (0.6 eV) is still not enough to resolve individual vibrational states, the shape of the profile reveals remarkably asymmetric, which is obviously the Franck-Condon profile arising from a series of vibrational excitations.

A least-squares fitting has been carried out utilizing a set of Gaussian functions for individual ionization transitions related to different vibrational states of H_2^+ in the BES. The width of each Gaussian function is fixed at present EMS instrumental energy resolution (0.6 eV) as the natural width of vibrational transition is negligible. The position of each Gaussian function is also referred to the exact energy position of the corresponding vibrational peak determined by high-resolution PES [31]. Thirteen Gaussian functions are used to identify the transitions related to the first thirteen vibrational states with vibrational quantum number $\mu' = 0 \sim 12$, respectively, and another one more function to fit the states with $\mu' > 12$. The experimental momentum profiles (XMPs), which are the momentum dependent EMS cross sections, can be extracted for individual vibrational states by plotting areas under the corresponding fitted peaks as a function of target electron momentum. The results are shown in figure 2(b)-(k). For $\mu' > 8$ only the summation is plotted in figure 2(k) due to the large uncertainty of data. As can be seen from the figures, all the XMPs show the typical *s*-type character due to the ionization from $1s\sigma_g$ of H_2. Also included in the figures for comparison are the theoretical momentum profiles (TMPs) calculated by

$$\frac{d^3\sigma}{d\Omega_a d\Omega_b dE_b} \propto \int \frac{d\Omega}{4\pi} |\langle pV'_{\mu'}i | 0V_0 \rangle|^2 \tag{10}$$

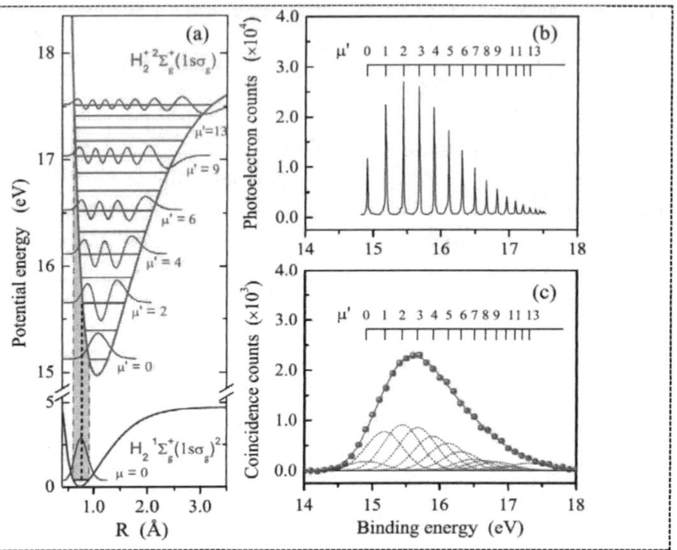

Figure 1. (a) The vibronic transitions from $X^1\Sigma_g^+$ ground state of H_2 to $^2\Sigma_g^+(1s\sigma_g)$ ground state of H_2^+. The wave functions of some vibrational states are also showed. (b) The high-resolution photoelectron spectrum of H_2. (c) The BES of H_2 obtained in present experiment. The dash lines indicate the Gaussian peaks, and the solid line is the summed fit.

For H_2 molecule which has only one vibrational mode, the structure amplitude $\langle pV'_{\mu'}i | 0V_0 \rangle$ can be simplified as

$$\langle pV'_{\mu'}i | 0V_0 \rangle = \int dR X^*_{\mu'}(R) X_0(R) S^{(i)}(R) \varphi(p, R) \tag{11}$$

where $S^{(i)}(R) = \int d\tau \phi_i(\tau, R) \psi(\tau, R)$ is the overlap integral of the wave functions for the final ion and molecular residue left after the knockout of an electron from molecular orbital, and $\varphi(p, R) = (1/2\pi)^{3/2} \int dr e^{-ir \cdot p} \psi(r, R)$ is the Fourier transform of $\psi(r, R)$. $X_0(R)$ and $X_{\mu'}(R)$ are vibrational wave functions of H_2 and H_2^+. The TMPs are convoluted with instrumental momentum resolution of 0.10 a.u. and all the TMPs and XMPs in the figures are area-normalized to unity. The agreements in shape between experiments and theoretical calculations are fairly good. Although the experimental uncertainties are rather large, the general trend that XMP becomes 'fatter' with increase of vibrational quantum number is somehow observable.

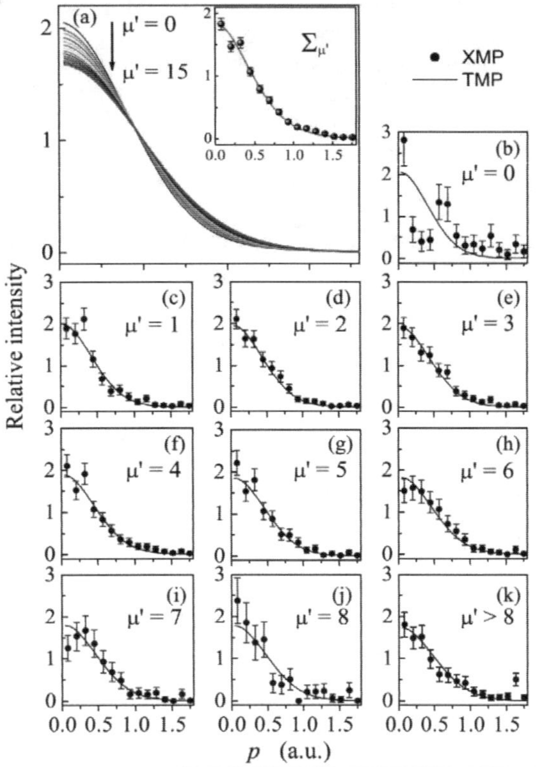

Figure 2. (a) The TMPs for different H_2^+ vibrational states (shown by different colors). The inset shows the summed XMP for all final vibrational states compared with the TMP calculated at equilibrium nuclear distance. The XMPs and corresponding TMPs for μ′ =0~8 are shown in (b)-(j), respectively. The XMP and TMP presented in (k) are the sum of XMPs and TMPs for μ′ >8. All profiles in this figure are area-normalized.

To highlight the differences, vibrational ratios of EMS cross sections are plotted in figure 3. We group the final vibrational states into three: the low quantum number group with $\mu' = 0,1,2$ denoted by capital letter *L*, the intermediate quantum number group with $\mu' = 4,5,6$ denoted by *I* and the high quantum number group with $\mu' = 9 \sim 13$ denoted by *H*. The theoretical EMS cross section $\sigma_{EMS}(15)$ for $\mu' = 15$ is chosen as a reference, and the summed EMS cross sections of both experiments and theoretical calculations for these three groups are all divided by the reference $\sigma_{EMS}(15)$. The ‘experimental’ and theoretical results of $\sigma_{EMS}(X)/\sigma_{EMS}(15)$ ($X = L, I, H$) are shown in figure 3(a)-(c) by solid circles and curves respectively. Here, $\sigma_{EMS}(X)$ are again area-normalized to unity respectively for the convenience of comparison. Using Franck-Condon (FC) approximation, equation (11) is reduced to

$$\langle pV'_{\mu'}i|0V_0\rangle = g_0^{\mu'} S^{(i)}(R_0)\varphi(p, R_0) \tag{12}$$

where $g_0^{\mu'} = \int dR X_{\mu'}(R) X_0(R)$ is the FC factor. Within this approximation, the ratio of EMS cross sections for two final vibrational states $\sigma_{EMS}(\mu_1')/\sigma_{EMS}(\mu_2')$ will be equal to the quotient of the relevant FC factor $g_0^{\mu_1'}/g_0^{\mu_2'}$, which will keep as a constant in the whole momentum range. However, one can see immediately from figure 3(a)-(c) that the observed ratios of $\sigma_{EMS}(X)/\sigma_{EMS}(15)$ ($X = L, I, H$) obviously deviates from constant (here is unity due to the normalization), especially the low quantum number group '*L*' whose ratio declines sharply with the momentum. The deviation from the FC principle can be ascribed to the two-center interference effect. When taking into account vibrational states, the vibrationally resolved cross section can be approximated by

$$\sigma_{EMS}(\mu') = \sigma_0 \left| \int_0^{\infty} X_{\mu'}(R) \left[1 + \frac{\sin(pR)}{pR} \right]^{1/2} X_0(R) dR \right|^2 \quad (13)$$

where σ_0 is the equivalent one center atomic cross section. By replacing in equation (13) the variable R by characteristic value $R_{\mu'}$ associated with μ'state, the vibrational ratio can further be approximated by

$$\frac{\sigma_{EMS}(\mu_1')}{\sigma_{EMS}(\mu_2')} = \frac{g_0^{\mu_1'}}{g_0^{\mu_2'}} \left[1 + \frac{\delta R_{\mu_1'}}{R_{\mu_2'}} \cos(pR_{\mu_2'}) \right] \quad (14)$$

This formula clearly predicts that the vibrational ratio should oscillate around the quotient of FC factor. To evaluate the observations, the turning points on potential curve for relevant vibrational states are adopted as the characteristic value $R_{\mu'}$ and the function $a_0 \left[1 + a_1 \frac{\delta R_{\mu_1'}}{R_{\mu_2'}} \cos(pR_{\mu_2'}) \right]$ is employed to fit the vibrational ratios, where a_0 and a_1 are adjustable parameters. The turning point for $\mu_2' = 15$ state is 1.12 a.u., while the averaged values of turning points 1.6 a.u., 1.35 a.u. and 1.15 a.u. are used for $X = L, I, H$ respectively. The parameter a_1 is introduced to compensate the approximations in the evaluations of $R_{\mu'}$ and thus is kept the same value for all three fittings. The fitted curves are presented in figure 3(a)-(c) as chain lines. The agreement of the model fitting with the measured and calculated vibrational ratios undoubtedly signifies the Young's two-center interference effect and the movement of the interference fringe has also been observed. Vibrationally resolved experiment provides a more straightforward way to observe Young-type interference in electron impact ionization of diatomic molecules, which does not rely on the comparison with one-center atomic cross section.

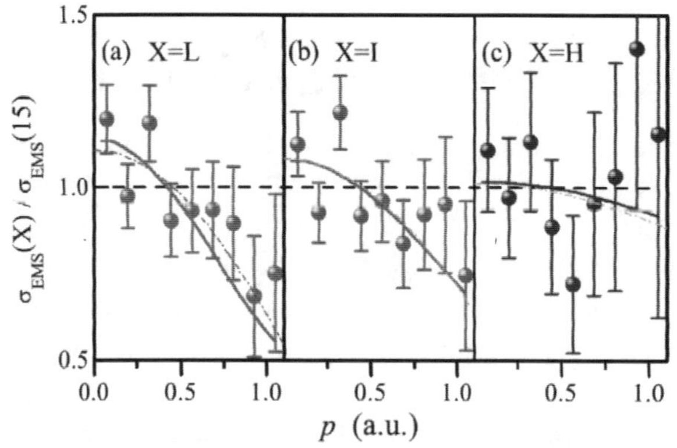

Figure 3. The experimental and theoretical ratios of the EMS cross sections for different groups of vibrational states over the calculated cross section for μ'=15 : (a) the low quantum number group $X = L$; (b) the intermediate quantum number group $X = I$; (c) the high quantum number group $X = H$. Solid circles and curves represent experimental and theoretical results. Dashed lines are Franck-Condon predictions. Chained lines are fitted curves of the model.

4. Multi-center interference effects in (e, 2e) of CF_4 and NF_3

As we have mentioned above, in order to have clear observations of interference effect, one way is to choose a molecule having larger internuclear distance. The multi-centre interference effect was observed in binary (e, 2e) experiment for the three outermost MOs of CF_4 [22] recently. The large F-F internuclear distance makes a complete period of oscillation to be observable. Most recently, a high-sensitivity angle and energy dispersive multichannel electron momentum spectrometer with simultaneous detection in 2π angle range has been developed [29]. The sensitivity of EMS has been improved by employing a double half wedge and strip anode (DH-WSA) position-sensitive detector (PSD) combined with a 90° sector, 2π spherical electrostatic analyzer. Furthermore, much wider momentum range from 0 to 8 a.u. has been achieved, which makes it possible to include more than two periods of oscillation in observations of multi-center interference effect in high symmetry molecules like NF_3 and CF_4. Here, only the results of CF_4 are presented.

In the simple MO theory, the ground state electronic configuration of CF_4 can be written as:

$$(\text{core})^{10}\ \underbrace{(3a_1)^2(2t_2)^6}_{\text{inner valence}}\ \underbrace{(4a_1)^2(3t_2)^6(1e)^4(4t_2)^6(1t_1)^6}_{\text{outer valence}}\ .$$

The three outermost orbitals are non-bonding, essentially due to the $2p$ lone-pair electrons on fluorine atoms. For these MOs consisting of the F 2p AOs, the TDCS can be expressed as [22]

$$\sigma_{\text{EMS}}(p) \propto \sigma_{2\text{p}}(p)\left[1 + C_0\, j_0(pR_{\text{FF}}) + C_2\, j_2(pR_{\text{FF}})\right] \tag{15}$$

where R_{FF} is the internuclear distance between the F atoms and C_0 and C_2 are coefficients of the spherical Bessel functions of order 0 and 2 respectively. The EMS cross section $\sigma_{2p}(p)$ for the $2p$ AO of an isolated F atom is calculated by distorted wave Born approximation employing B3LYP/aug-cc-pVTZ wavefunction. The function $[1 + C_0 j_0(pR_{FF}) + C_2 j_2(pR_{FF})]$ governs the oscillatory structure and hence it is the interference factor in this case

$$\sigma_{\text{EMS}}(p)/\sigma_{2\text{p}}(p) \propto 1 + C_0\, j_0(pR_{\text{FF}}) + C_2\, j_2(pR_{\text{FF}}). \tag{16}$$

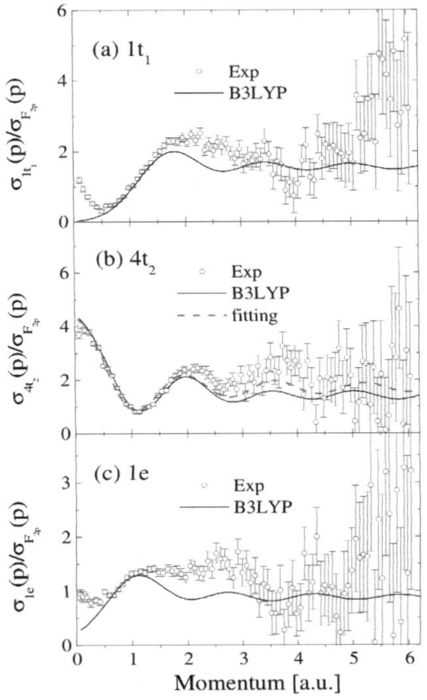

Figure 4. Comparison of experimental and DWBA/B3LYP calculated interference factor for the (a) $1t_1$, (b) $4t_2$, and (c) $1e$ molecular orbitals of CF_4.

The interference factors for the three outermost orbitals of CF_4 are shown in figure 4. It is immediately clear that the experiments exhibit oscillatory structures. Compared to the previous experiment [22], more periods of oscillations are included, further confirming the interference effect. To highlight it more closely, the function $h[1 + C_0 j_0(pR_{FF}) + C_2 j_2(pR_{FF})]$ is subsequently employed as a fitting curve for $4t_2$ to reproduce the experiment with R_{FF}, h, C_0, and C_2 being fitting parameters. The best fit to the experiment is presented also in figure 4(b) by the dashed line. The resulting R_{FF} value is 3.9 Bohr which is in excellent agreement with 4.07 Bohr reported by electron diffraction [32].

Acknowledgments
This work was jointly supported by the National Natural Science Foundation of China (Grant Nos. 11327404, 10734040), the National Basic Research Program of China (Grant No. 2010CB923301).

References

[1] Jönsson C 1961 *Z. Phys.* **161** 454
[2] Arndt M, Nairz O, Vos-Andreae J, Keller C, van der Zouw G and Zeilinger A 1999 *Nature* **401** 680
[3] Gerlich S, Eibenberger S, Tomandl M, Nimmrichter S, Hornberger K, Fagan P J, Tüxen J, Mayor M and Arndt M 2011 *Nature Comunications* **2** 263
[4] Cohen D and Fano U 1966 *Phys. Rev.* **150** 30
[5] Stolterfoht N *et al.* 2001 *Phys. Rev. Lett.* **87** 23201
[6] Misra D, Kadhane U, Singh Y P *et al.* 2004 *Phys. Rev. Lett.* **92** 153201
[7] Schmidt H T, Fischer D, Berenyi Z *et al.* 2008 *Phys. Rev. Lett.* **101** 083201
[8] Misra D, Schmidt H T, Gudmundsson M *et al.* 2009 *Phys. Rev. Lett.* **102** 153201
[9] Voitkiv A B, Najjari B and Fischer D 2011 *Phys. Rev. Lett.* **106** 233202
[10] Rolles D, Braune M, Cvejanović S *et al.* 2005 *Nature* **437** 711
[11] Fernández J, Fojón O, Palacios A and Martín F 2007 *Phys. Rev. Lett.* **98** 043005
[12] Akoury D, Kreidi K, Jahnke T *et al.* 2007 *Science* **318** 949
[13] Kreidi K, Akoury D, Jahnke T *et al.* 2008 *Phys. Rev. Lett.* **100** 133005
[14] Zimmermann B, Rolles D, Langer B *et al.* 2008 *Nature Physics* **4** 649
[15] Canton S E, Plésiat E, Bozek J D *et al.* 2011 *PNAS* **108** 7302
[16] Kushawaha R K, Patanen M, Guillemin R *et al.* 2013 *PNAS* **110** 15201
[17] Stia C R, Fojón O A, Weck P F *et al.* 2003 *J. Phys. B* **36** L257
[18] Milne-Brownlie D S, Foster M, Gao J *et al.* 2006 *Phys. Rev. Lett.* **96** 233201
[19] Staicu-Casagrande E M *et al.* 2008 *J. Phys. B* **41** 025204
[20] Kamalou O, Chesnel J Y, Martina D *et al.* 2005 *Phys. Rev. A* **71** 010702(R)
[21] Ozer Z N, Chaluvadi H, Ulu M *et al.* 2013 *Phys. Rev. A* **87** 042704
[22] Watanabe N, Chen X J and Takahashi M 2012 *Phys. Rev. Lett.* **108** 173201
[23] McCarthy I E and Weigold E 1991 *Rep. Prog. Phys.* **54** 789.
[24] Epstein I R and Tanner A C, in *Compton Scattering*, edited by B. G. Williams (McGraw-Hill, New York, 1977)
[25] Cook J P D and Brion C E 1982 *Chem. Phys.* **69** 339
[26] Coulson C A 1941 *Proc. Cambridge Philos. Soc.* **37** 55;
Coulson C A and Duncanson W E 1941 *Proc. Cambridge Philos. Soc.* **37** 67
[27] Zhang Z, Shan X, Wang T *et al.* 2014 *Phys. Rev. Lett.***112** 023204
[28] Shan X, Chen X J, Zhou L X *et al.* 2006 J. Chem. Phys. **125**, 154307 (2006)
[29] Tian Q G, Wang K D, Shan X and Chen X J 2011 *Rev. Sci. Instrum.* **82** 033110
[30] Dey S, McCarthy I E, Teubner P and Weigold E 1975 *Phys. Rev. Lett.* **34** 782
[31] Pollard J E 1982 *J. Chem. Phys.* **77** 34
[32] Fink M, Schmiedekamp C W and Gregory D 1979 *J. Chem. Phys.* **71** 5238

Journal of Physics: Conference Series **601** (2015) 012003 doi:10.1088/1742-6596/601/1/012003

Interference effects for intermediate energy electron-impact ionization of H_2 and N_2 molecules

Zehra Nur Ozer[1], Hari Chaluvadi[2], Don Madison[2] and Mevlut Dogan[1]
[1]Department of Physics, e-COL Laboratory, Afyon Kocatepe University, 03200, Afyonkarahisar, Turkey
[2]Department of Physics, Missouri University of Science and Technology, Rolla, Missouri, USA

E-mail: zehraerengil@aku.edu.tr

Abstract. We have studied electron impact ionization of H_2 and N_2 molecules at intermediate energies to look for possible two center interference effects experimentally and theoretically. Here we report a study of the interference factor I for 250 eV electron-impact ionization. The experimental measurements are performed using a crossed-beam-type electron-electron coincidence spectrometer and theoretical calculations are obtained using the Molecular Three Body Distorted Wave Approximation (M3DW). We found that the I-factor demonstrated strong evidence for two-center interference effects for both H_2 and N_2. We also found that the I-factor is more sensitive to projectile angular scans than to ejected electron energy scans which indicate that for the present set of kinematics the diffraction of the projectile from two scattering centers is more important than interference between electron waves emitted from two different centers.

1. Introduction

Electron impact ionization cross sections have been measured and calculated since the early days of collision physics due to the wealth of information that can be obtained about the collision dynamics and also due to their relevance in many application areas. Because of the importance of the cross sections in biological applications, much emphasis has been devoted recently to the experimental studies of electron-impact ionization cross sections of molecules and radicals. The study of the energy and angular distributions of electrons ejected by electron impact is a sensitive means of testing the theory of collision processes.

The possibility of a diatomic molecule acting as a microscopic double slit for photon impact was suggested by Cohen and Fano several decades ago [1]. Since quantum particles are also waves, the next natural question was whether Young's type interference effects could also be observable in triple differential cross section (TDCS) spectra for electron-impact ionization of diatomic molecules. Young type interference effects resulting from the coherent superposition of the scattered waves from two atomic centers were predicted by Stia et al. [2-3] and by Gao et al. [4] for electron impact ionization of molecular H_2 and molecular N_2, respectively. However, due to the small cross sections, the experimental investigations of the ionization of small molecules by this technique are limited. There have been two experimental studies, in coplanar asymmetric geometry, presented for electron impact ionization of H_2 (Milne-Brownlie et al. in 2006 [5] and Casagrande et al. in 2008 [6]). These studies found evidence for interference effects by comparing the relative sizes of the binary and recoil peaks.

The theoretical basis for these studies lies in the interference factor which is defined to be the ratio of the molecular cross section divided by the cross section for two atoms. The idea is that dividing by the atomic cross section should isolate the molecular two-center effects. Cohen and Fano [1] showed that, for photon impact on H_2, this factor could be approximated as

$$I^{CF} = 1 + \frac{\sin(\chi \rho_0)}{\chi \rho_0}$$

where ρ_0 denotes the equilibrium internuclear vector of the molecular target and χ is the momentum transfer. Stia et al. [2-3] showed that the same approximation could be used for electron impact ionization of H_2. This factor predicts that the recoil peaks for atomic H should be either larger or smaller than molecular H_2 depending on the kinematics. Due to the difficulty in measuring atomic H cross sections, the experimentalists substituted He for two atomic H and found the predicted enhancement/suppression in the recoil peak.

Gao et al. [4] predicted similar interference effects for ionization of the $3\sigma_g$ state of N_2. Murray et al. [7] and Hargreaves et al. [8] performed experiments for ionization of N_2 for both asymmetric and symmetric geometries. However, their results did not provide strong evidence for these interference effects.

There are three different types of two center interference effects for electron-impact scattering: (1) Incident electron being diffracted by two scattering centers; (2) scattered electron wave being emitted from two centers; and (3) ejected electron wave being emitted from two centers. We performed calculations for the three different types of possible interference effects for ionization of H_2 and we found that the most important contribution comes from the incident projectile diffracting from two scattering centers [9-10]. As an extension of this study, we have now examined electron impact ionization of N_2 molecules at intermediate energies to look for interference effects both experimentally and theoretically. The experimental measurements are performed using a crossed-beam-type electron-electron coincidence spectrometer and theoretical calculations are obtained using the Molecular Three Body Distorted Wave Approximation (M3DW) [11].

2. Experimental Apparatus

The measurements were performed at e-COL laboratory (Afyon, Turkey) using an electron spectrometer that is designed for electron-electron coincidence (e,2e) experiments. The details of the electron spectrometer are described in detail in previous papers [9, 12-14]. Briefly, the electron spectrometer consists of a electron gun producing a beam of electrons which passes through the gas target perpendicularly, two hemispherical electrostatic energy analyzers, a Faraday cup and a data acquisition system (see in figure 1a). The spectrometer is contained in a cylindrical stainless steel vacuum chamber and the pressure in the chamber is around ~2×10^{-6} mbar while the experiment is running. The spectrometer operated at an electron current of ~2 μA with a resolution of ~0.7 eV. The (e,2e) technique is used to detect two outgoing electrons in coincidence after the ionization of the target. Two electrons of the desired energies are detected and amplified using Channel Electron Multipliers (CEM). This technique has an advantage of obtaining single ionization events meaning the outgoing electrons have originated from the same ionization event. To do this, time correlation between the detected electrons are taken into consideration and time delay between the electrons is converted to a signal that is measured by computer, and a narrow coincidence peak in the timing spectrum is observed. Coincidence electronics are shown in figure 1b.

The overall energy resolution of the coincidence system was limited by both the thermal spread of the electrons emitted from the tungsten hairpin cathode and the analyzer system. By measuring the full width at half-maximum (FWHM) of the binding energy spectra, the coincidence energy resolution was found to be ≈1.5 eV. This resolution is good enough to separate the $3\sigma_g$ orbital of the N_2 molecule.

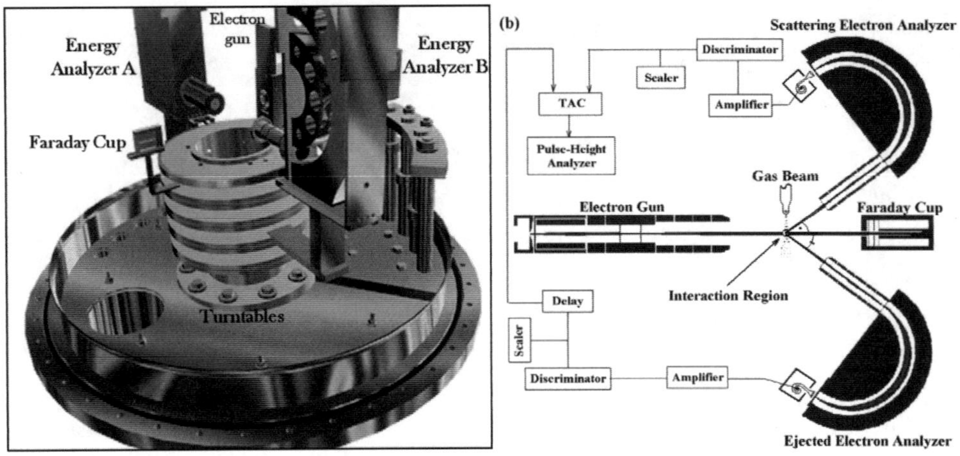

Figure 1. Sketch of electron spectrometer. The main components are electron gun, two electron analyzers and Faraday cup that are mounted on the turntables (a). Coincidence electronics used to accumulating a coincidence timing spectrum at each kinematics (b).

3. Theory

Al-Hagan et al. [11] showed that the molecular three-body distorted wave approximation (M3DW) coupled with an orientation-averaged molecular orbital approximation [15] yielded a good agreement with experimental TDCS data for H_2. The molecular 3-body distorted wave (M3DW) approximation has been presented elsewhere [16-17] so only a brief description of the theory will be presented. The triple differential cross section (TDCS) for the M3DW is given by:

$$\frac{d\sigma}{d\Omega_a d\Omega_b dE_b} = \frac{1}{(2\pi)^5}\frac{k_a k_b}{k_i}\left(\left|T_{dir}\right|^2 + \left|T_{exc}\right|^2 + \left|T_{dir} - T_{exc}\right|^2\right) \tag{1}$$

where $\vec{k}_i$, $\vec{k}_a$, and $\vec{k}_b$ are the wave vectors for the initial, scattered and ejected electrons, and T_{dir} is the direct scattering amplitude given by:

$$T_{dir} = \left\langle \chi_a^-(\vec{k}_a,\mathbf{r}_1)\chi_b^-(\vec{k}_b,\mathbf{r}_2)C_{scat-eject} \,|\, V - U_i \,|\, \phi_{DY}^{OA}(\mathbf{r}_2)\chi_i^+(\vec{k}_i,\mathbf{r}_1)\right\rangle \tag{2}$$

where r_1 and r_2 are the coordinates of the incident and the bound electrons, χ_i, χ_a, and χ_b are the distorted waves for the incident, scattered, and ejected electrons respectively, and $\phi_{DY}^{OA}(r_2)$ is the initial bound-state Dyson molecular orbital averaged over all orientations. The factor $C_{scat-eject}$ is the Coulomb interaction between the two final state electrons. We have found that using the exact interaction can over-estimate the strength of the interaction for atoms so we have used the Ward-Macek approximation [18] for atoms and the exact interaction for molecules. The initial state interaction potential between the incident electron and the neutral molecule is V, and U_i is a spherically symmetric approximation for V. This potential is called the distorting potential and it is used to calculate the initial-state distorted wave for the incident electron $\chi_i^+(\vec{k}_i,\mathbf{r}_1)$. For the exchange amplitude T_{exc}, particles 1 and 2 are interchanged in equation (2). The Schrödinger equation for the incoming electron wave-function is given by:

$$(T + U_i - \frac{k_i^2}{2})\chi_i^+(\vec{k}_i, r) = 0 \tag{3}$$

where T is the kinetic energy operator and the '+' superscript on $\chi_i^+(\vec{k}_i, \mathbf{r})$ indicates outgoing wave boundary conditions. The initial state distorting potential contains three components $U_i = U_s + U_E + U_{CP}$, where U_s contains the nuclear contribution plus a spherically symmetric approximation for the interaction between the projectile electron and the target electrons which is obtained from the quantum mechanical charge density of the target. For H_2, the charge density is $2|\phi_{DY}|^2$ (the 2 is for double occupancy and the original non-averaged Dyson orbital is used). For N_2, the charge density is summed over the seven occupied orbitals. The nuclear contribution to U_s is the interaction between the projectile electron and the 2 nuclei averaged over all orientations which means that the net charge of the two nuclei is placed on a thin shell whose radius is the distance of the nuclei from the center-of-mass.

The potential U_E is the exchange potential of Furness-McCarthy [19] which approximates the effect of the continuum electron exchanging with the passive bound electrons in the molecule, and U_{CP} is the correlation-polarization potential of Perdew and Zunger [20] (see also Padial and Norcross [21]).

The final state distorted waves are obtained the same as the initial state except that the final state charge density for the ion is used to calculate U_s. The final state charge density is obtained the same as the initial state except that the occupancy number for the active electron is unity.

4. Results and Discussion

In a previous paper, we examined the TDCS for electron-impact ionization of the H_2 molecule in comparison to atomic He. We found that there was an overall good agreement between experiment and theory [9]. Both experiment and theory predict a much more complicated interference pattern, particularly in the binary peak region, than is given by the elementary Cohen-Fano interference factor, I^{CF}. We also found that the I-factor is more sensitive to projectile angular scans than to ejected-electron energy scans which indicate that for the present set of kinematics the diffraction of the projectile from two scattering centers is more important than the interference between electron waves emitted from two different centers.

In this work, we have extended our TDCS measurements to N_2 molecules for the same kinematics. Figure 2 compares H_2 and N_2 results for 250 eV incident electron impact, fast electron scattering angle of 15° and ejected electron of 50 eV. The top two panels of the figure contain the previously presented results for H_2 and He. On the left are the TDCS for both the atom and molecule normalized to unity at the TDCS peak and the right panel contains the I-factors. The experimental I-factor is the ratio of the H_2 and He (not shown) TDCS on the left, the red solid curve on the right is the ratio (I^{M3DW}) of the M3DW H_2 and He TDCS shown on the left and the blue dashed curve is I^{CF} (also sometimes called the Stia factor).

While the elementary Cohen-Fano factor exhibits qualitatively the shape of the I-factor, both the experimental and M3DW I-factors indicate a more complicated structure and the M3DW is in reasonable qualitative agreement with experiment. Results for N_2 and N are presented in the bottom 2 panels. The left panel presents theoretical and experimental results for N_2 as well as theoretical results for atomic N. All the TDCS have been normalized to unity at the peak. The lower right panel contains the corresponding I-factors. Since there were no experimental TDCS available for atomic N, the experimental I-factors for N_2 represent the experimental N_2 TDCS divided by the theoretical atomic N TDCS. Again it is seen that the experimental and M3DW I-factors have a more complicated structure than I^{CF}. While the I-factors for H_2 and N_2 are similar, the I-factor for H_2 exhibits a double peak structure within the angular range of binary peak which is supported by the experimental data while N_2 has only a single peak in this angular range with a much smaller peak outside the angular range of the experiment. The M3DW I-factor for N_2 is in very good agreement with experiment for all the measured points.

The logic behind the I-factor is that dividing the molecular TDCS by the atomic TDCS removes all atomic effects in the TDCS and leaves two-center (molecular) effects. The Cohen-Fano I^{CF} factor shows what this interference is predicted to look like for photon impact where the interference would result from the target electron being ejected from two centers. The fact that the experiment and theory qualitatively have the shape of the Cohen-Fano factor indicates that this type of interference is present. The fact that the actual structure is much more complicated indicates that the other two possible modes of interference (i.e. incident electron being diffracted from two scattering centers and the scattered electron wave being emitted from two centers) is more important.

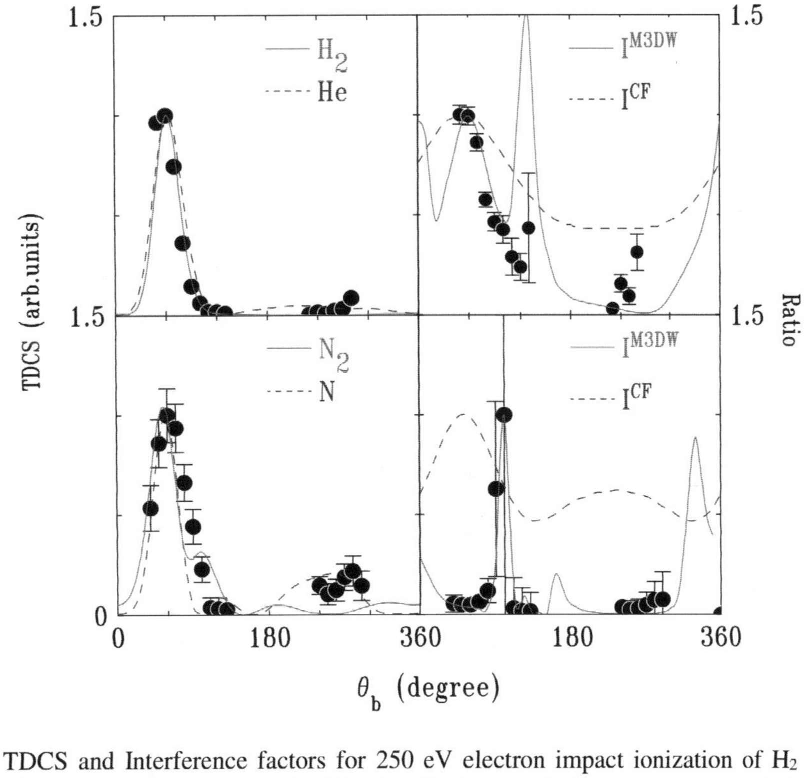

Figure 2. TDCS and Interference factors for 250 eV electron impact ionization of H_2 and N_2 as a function of the ejected electron angle θ_b. The fast final state electron scattering angle is 15° and the ejected electron energy is 50 eV. The top two panels of the figure contain results for H_2 and He (TDCS on left and TDCS ratios on right) and the bottom two panels contain results for N_2 and N. Solid circles-present data; solid red curve on the left - M3DW TDCS for H_2 and N_2; dashed blue curve on the left - M3DW TDCS for He and N; solid red curve on the right - I^{M3DW}; dashed blue curve on right I^{CF}. See text for more information.

In summary, we see that there is significant interference at the quantum level that it is not amenable to a simple classical interpretation for lower energy incident electrons. These results demonstrate for the first time that Young's type double slit interference effects are present for ionization of N_2 as well as H_2. These are preliminary results. We are in the process of measuring more TDCS for N_2 at other projectile scattering angles and ejected electron energies and will present a more complete set of results in the near future.

Acknowledgments
This work was supported by the Scientific and Technological Research Council of Turkey (TUBITAK) through grant 109T738. The theoretical work of HC and DM was supported by US National Science Foundation under Grant. No. PHY-1068237.

References
[1] Cohen H D and Fano U 1966 *Phys. Rev.* **150** 30
[2] Stia C R, Fojon A O, Weck P F, Hanssen J and Rivarola R 2003 *J. Phys. B.* **36** L257
[3] Stia C R, Fojon A O, Weck P F, Hanssen J, Joulakian B and Rivarola R 2002 *Phys. Rev. A* **66** 052709
[4] Gao J, Madison D H and Peacher J L 2005 *Phys. Rev.* A **72** 032721
[5] Milne-Brownlie D A, Foster M, Gao J, Lohmann B and Madison D H 2006 *Phys. Rev. Lett.* **96** 233201
[6] Casagrande E M, Lahmam-Bennani A and Madison D H 2008 *J. Phys. B* **41** 025204
[7] Murray A, Hussey M J, Gao J and Madison D H 2006 *J. Phys. B* **39** 3945
[8] Hargreaves et al. 2009 *Phys. Rev. A* **80** 062704
[9] Ozer Z N, Chaluvadi H, Ulu M, Dogan M, Aktas B and Madison D 2013 *Phys. Rev. A* **87** 042704
[10] Ozer Z N, Chaluvadi H, Ulu M, Dogan M, Aktas B and Madison D 2014 *J. Phys.: Conf. Ser.* **488** 012059
[11] Al-Hagan O, Kaiser C, Madison D and Murray A 2009 *Nature Physics* **5** 59
[12] Dogan M, Ulu M and Sise O 2007 *J. Elect. Spect. Rel. Phen.* **161** 58-62
[13] Sise O, Dogan M, Okur I and Crowe A 2011 *Phys. Rev. A* **84** 022705
[14] Ozer Z N, Ulu M, Aktas B, Dogan M 2013 *Acta Physica Polonica A* **123 2** 363-364
[15] Madison D H and Al-Hagan O, *Journal of Atomic, Molecular, and Optical Physics* 2010 367180
[16] Gao J, Madison D H and Peacher J L 2005 *J. Chem. Phys.* **123** 204314
[17] Gao J, Madison D H and Peacher J L 2005 *Chem. Phys.* **123** 204302
[18] Ward S J and Macek J H 1994 *Phys. Rev. A* **49** 1049
[19] Furness J B and McCarthy I E 1973 *J. Phys. B* **6** 2280
[20] Perdew J P and Zunger A 1981 *Phys. Rev. B* **23** 5048
[21] Padial N T and Norcross D W 1984 *Phys. Rev. A* **29** 1742

Generalized Sturmian Functions applied to double continuum problems

M. J. Ambrosio[1,5], F. D. Colavecchia[2,5], D. M. Mitnik[1,5], G. Gasaneo[3,5] and L. U. Ancarani[4]

[1] Instituto de Astronomía y Física del Espacio, Argentina.
[2] División Física Atómica, Molecular y Óptica, Centro Atómico Bariloche, 8400 S. C. de Bariloche, Río Negro, Argentina.
[3] Departamento de Física, Universidad Nacional del Sur, 8000 Bahía Blanca, Buenos Aires, Argentina.
[4]Théorie, Modélisation, Simulation, SRSMC, UMR CNRS 7565, Université de Lorraine, 57078 Metz, France
[5] Consejo Nacional de Investigaciones Científicas y Técnicas, CONICET, Argentina.

E-mail: mja1984@gmail.com

Abstract. The Generalized Sturmian Functions method aims to deal with atomic physics problems. It has seen application to two and three–body problems, and its flexibility enables one to work with bound systems as well as with particles in the continuum. In the present contribution we analyze how the method expands the atomic double continuum in collision problems, using the double ionization of Helium by fast electrons as a showcase. We first test the robustness of the method in a particularly challenging situation, the zero energy case. We then present fully differential cross sections for a scattering problem which after 15 years of continued efforts has not been satisfactorily solved: the double ionization of Helium by electron impact in the fast projectile regime, as measured by the Orsay group.

1. Introduction

The Generalized Sturmian Functions (GSF) methodology was introduced seven years ago [1], and has seen a series of successful applications in atomic physics problems (see [2] and references therein). Whether in the case of bound systems or particles in the continuum, the method has provided accurate solutions at a computationally inexpensive cost (see, e.g., Ref. [3]). GSF form basis sets which have two main characteristics underlying their efficiency. First, the basis functions can be constructed in such a way to contain the exact or approximate asymptotic conditions for a given problem. Second, the oscillations of the basis functions can be localized in any spatial region of interest, leaving the expansion to deal essentially with the domain where it is most needed.

For the case of bound systems, it is easy to understand how the basis functions provide excellent convergence properties, concentrating the expansion capabilities in the (inner) region of interest [4, 5, 6, 7] and having all basis elements decay exponentially at the same rate, as close as possible to the expected behavior of the physical wave function.

Less clearly established is how the basis manages to solve three–body double continuum problems, although the efficiency has been illustrated through application to the study of single electron impact ionization of hydrogen [3, 8]. This contribution aims to analyze how the hyperspherical asymptotic behavior for the double continuum wave fronts can be well approximated by the GSF method, by using an angularly coupled combination of basis functions in the radial coordinates of the two emitted particles.

The proposed analysis is done here for the double ionization of Helium in the fast projectile regime. This process is formally a four–body problem, but the collision dynamics can be reduced to a three–body problem when treating the projectile in a perturbative fashion [9]. For the purpose of the present investigation, in the first part we restrict ourselves to a Temkin–Poet (TP) model of the physical process. While keeping the main challenges of a three–body problem, it allows to perform studies on a single partial wave. Then, going beyond the TP case, the full three–body problem is studied and Fivefold Differential Cross Sections (FDCS) are briefly presented. We also provide some convergence details related to the relevance of angular momentum partial waves.

Atomic units ($\hbar = e = m_e = 1$) are assumed throughout, unless stated otherwise.

2. Double ionization of Helium by fast electrons: GSF expansion

The double ionization of Helium formalism used by our group was introduced in [9]. The system is comprised of the nucleus ($Z = 2$), the target electrons and the projectile. With the same notation as in Ref. [9, 10, 11], r_2 and r_3 denote the distances between the nucleus and the target electrons, and r_{23} the interelectronic distance. Vector $\mathbf{q}$ stands for the momentum transferred from the projectile to the originally bound system; in this work we shall take $q = 0.24$ a.u. which corresponds to one of two configurations explored by the Orsay group in Refs. [12, 13, 14]. The Helium target is left with total energy E_a after the collision. The formulation treats the projectile–He interactions in a perturbative fashion [9]; considered up to its first order, it yields the following equation for the ejected electrons' scattering wave function:

$$[h_{He} - E_a]\,\Phi^+_{sc}(\mathbf{q},\mathbf{r}_2,\mathbf{r}_3) = -\frac{4\pi}{q^2}\frac{1}{(2\pi)^3}\left(-Z + e^{i\mathbf{q}\cdot\mathbf{r}_2} + e^{i\mathbf{q}\cdot\mathbf{r}_3}\right)\Phi_i(\mathbf{r}_2,\mathbf{r}_3), \tag{1}$$

where the three–body Hamiltonian h_{He} is given by

$$h_{He} = \left(-\frac{1}{2}\nabla^2_2 - \frac{1}{2}\nabla^2_3 - \frac{Z}{r_2} - \frac{Z}{r_3} + \frac{1}{r_{23}}\right). \tag{2}$$

The Hamiltonian appearing in Eq. (1) is separable in the total angular momentum subspaces, implying that the well known bipolar harmonics $\mathcal{Y}^{LM}_{l_2l_3}(\widehat{\mathbf{r}}_2,\widehat{\mathbf{r}}_3)$ are eigenfunctions of this operator. Consequently, we construct the three–body GSF basis $\Theta_\nu(\mathbf{r}_2,\mathbf{r}_3)$ as the non–correlated product (i.e., with no explicit r_{23} dependence)

$$\Theta_\nu(\mathbf{r}_2,\mathbf{r}_3) = \mathcal{Y}^{LM}_{l_2l_3}(\widehat{\mathbf{r}}_2,\widehat{\mathbf{r}}_3)\,\frac{S_{n_2l_2}(r_2)}{r_2}\frac{S_{n_3l_3}(r_3)}{r_3}, \tag{3}$$

where index ν denotes the set of indices $\{L, M, l_2, l_3, n_2, n_3\}$. For three–body problems, the GSF method expands the continuum solution $\Phi^+_{sc}(\mathbf{r}_3,\mathbf{r}_3)$ in terms of this basis set:

$$\Phi^+_{sc}(\mathbf{q},\mathbf{r}_2,\mathbf{r}_3) = \sum_{L,M}\sum_{l_a,l_b}\sum_{n_a,n_b}\phi_\nu\Theta_\nu(\mathbf{r}_2,\mathbf{r}_3), \tag{4}$$

with coefficients ϕ_ν. The two–body functions $S_{n_2l_2}(r_2)$ and $S_{n_3l_3}(r_3)$ in Eq. (3) are the so–called Generalized Sturmian Functions, and verify a homogeneous differential equation which resembles the reduced radial Schrödinger equation:

$$[\mathcal{T}_l + \mathcal{U}(r) - E_s]\,S_{nl}(r) = -\beta_{nl}\,\mathcal{V}(r)\,S_{nl}(r), \tag{5}$$

with $\mathcal{T}_l = -\frac{1}{2\mu}\frac{d^2}{dr^2} + \frac{l(l+1)}{2\mu r^2}$, μ the reduced mass (set to 1 throughout this paper) and l the angular momentum quantum number. The energy E_s is fixed externally, and in principle, is an arbitrary parameter. A good choice of E_s, however, plays a key role to describe the double continuum behavior, as will be discussed later. The potential $\mathcal{U}(r)$ is used to include in the basis set some physics of the problem in question: mainly to account for non–active electrons shielding the nuclear charge and for the asymptotic charge perceived at long ranges by the active electrons. The other potential involved in Eq. (5), $\mathcal{V}(r)$, is called *generating potential*, and its role is to concentrate the oscillations of the basis in any desired region, thus tailoring the set for any specific application.

Sturmians with negative E_s decay exponentially and are suitable for bound state studies. For positive E_s, the asymptotic behavior can be chosen as a linear combination of pure Coulombic waves $H_l^+(r)$ and $H_l^-(r)$. For continuum applications we choose $E_s > 0$ and $S_{nl}(r) \rightarrow H_l^+(r)$ beyond a given radius R. Past the point $r = R$, all basis elements become linearly dependent (LD) and exhibit the same asymptotic behavior.

As it has been shown in previous publications (see [2] and references therein), the GSF method is particularly efficient resource wise when dealing with three–body bound states, and also with the three–body continuum [3]. In this contribution it is our wish to further investigate why the methodology works well in building up the double continuum. The main point is the following. The GSF basis imposes outgoing (+) type asymptotic conditions on each coordinate r_2 and r_3, i.e., on a square contour in the (r_2, r_3) space. However, the double continuum asymptotic behavior comprises outgoing waves in the hyperradial coordinate $\rho = \sqrt{r_2^2 + r_3^2}$, i.e., on a circular contour. We will see below that, with an adequate choice of energy E_s for the basis set, the outgoing hyperspherical condition is very well approximated by the GSF implementation in spherical coordinates (r_2, r_3).

3. Double continuum asymptotic conditions

Ideally, for a double continuum wave function to be a proper solution of the three-body problem, the asymptotic conditions imposed by a resolution scheme should contain or approximate the hyperspherical behavior formally established by Peterkop [15] and Kadyrov [16, 17]. Thus, the pure hyperspherical double continuum should be imposed in order to perfectly expand the expected solution. This would imply a scattering function with an asymptotic wave vector given by $\boldsymbol{\kappa} = \kappa\widehat{\boldsymbol{\rho}}$, defining for the purpose $\widehat{\boldsymbol{\rho}} = \cos(\alpha)\ \widehat{\mathbf{e}}_2 + \sin(\alpha)\ \widehat{\mathbf{e}}_3$, $\kappa = \sqrt{2E_a}$, and where $\tan(\alpha) = r_3/r_2$. Unit vectors $\widehat{\mathbf{e}}_2$ and $\widehat{\mathbf{e}}_3$ point in the (r_2, r_3) space, respectively along axes r_2 and r_3.

In figure 1 we show the scattering wave function $\Phi_{sc}^+(\mathbf{q}, \mathbf{r}_2, \mathbf{r}_3)$, calculated within a RxR=60x60 a.u. box, but plotted beyond the range of the calculation, where the basis sets are no longer linearly independent (LI). Sector I is where the basis is LI in both coordinates. In sector II we are evaluating the function beyond the $r_2 = R$ limit, where the set $S_{n_2l_2}(r_2)$ has become LD and $S_{n_3l_3}(r_3)$ still has expanding capabilities (i.e., is LI); in sector III we have the symmetric situation. Sector IV shows $\Phi_{sc}^+(\mathbf{q}, \mathbf{r}_2, \mathbf{r}_3)$ plotted when the basis is, in both coordinates, outside of its LI range. It contains pure outgoing wave fronts. We stress here that the calculation is performed inside the box, and physical results (cross sections) are always extracted from within it.

Consider the transition from sector I to II. For $r_2 > R$, the GSF set $S_{n_2l_2}(r_2)$ behaves as an outgoing wave in coordinate r_2, becoming LD. But we are still inside the r_3 range where the elements $S_{n_3l_3}(r_3)$ are LI. Therefore, in the r_3 direction, the basis can expand the behavior of the expected hyperspherical front, specifically, an outgoing wave on r_3 with wave number $\kappa\sin(\alpha)$. It is then in the r_2 part that there is an approximation: the component of $\boldsymbol{\kappa}$ across the boundary is $\kappa\cos(\alpha)$ while the set $S_{n_2l_2}(r_2)$ approximates it by κ. The degree of approximation equates to that of using $\cos(\alpha) \approx 1$, carrying an error term of the order $\alpha^2/2$. This is the reason why the particular choice $E_s = E_a$ manages, by imposing asymptotic conditions in coordinates (r_2, r_3), to approximate very well the expected hyperspherical structure. Analogously, by symmetry, the same logic applies when crossing from I to III.

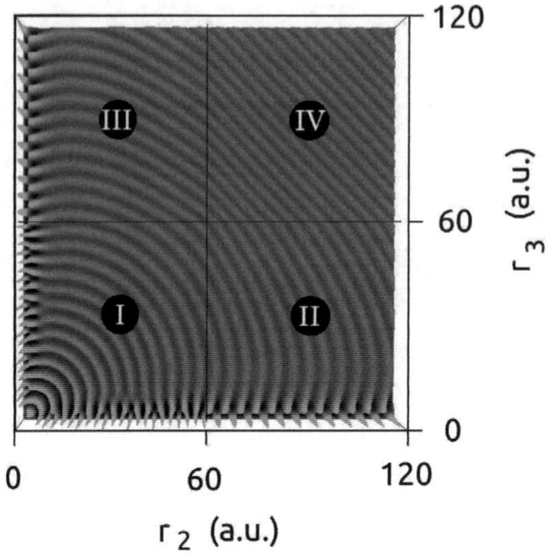

Figure 1. (Color online) Scattering wave function (real part) plotted as a function of r_2 and r_3. The calculation was performed inside a 60x60 a.u. box, but plotted in a 120x120 a.u. box. In sector II (III) the basis becomes LD in coordinate r_2 (r_3), but not in r_3 (r_2). In sector IV, the basis is LD in both coordinates.

These arguments justify the empirical affirmation made in Ref. [3], where it is suggested that taking $E_s = E_a$ provides the most precise three–body wave functions within the GSF method. We found that this choice is enough to ensure sound results to extract the transition amplitude from a region limited by $r_2, r_3 \approx 0.9\ R$, where the wave function is considerably neat and devoid of noise.

Regarding the hyperspherical double continuum, it is worth noting that hyperspherical GSF are being developed [18, 19]. In fact, coincident results for the double ionization of Helium by fast electrons with the spherical and hyperspherical GSF were jointly presented in [9]. It was suggested in [10, 20] that both versions should be mutually complementary in the sense that the single continuum information contained in the three–body wave function is better characterized with a spherical basis set, while the hyperspherical version is expected to be more efficient handling the double continuum. The hyperspherical GSF implementation is currently being generalized to go beyond the S-wave frame.

Having explained how the basis works to approximate the hyperspherical outgoing boundary conditions, we show the GSF method's consistency through a challenging test. It is very demanding for any method to solve Eq. (1) in the zero energy case ($E_a = 0$), which classically implies that both Helium electrons barely have the energy to be detached from the core. They distance themselves from the nucleus with a vanishing asymptotic velocity. This is reflected in the quantum mechanical asymptotic function described by Klar [21], with a wavelength growing with ρ:

$$\Phi_{sc}^{+}(\mathbf{q}, \mathbf{r}_2, \mathbf{r}_3) \rightarrow \frac{1}{\rho^{\frac{9}{4}}} Exp\left[A(\alpha, \theta)\sqrt{\rho}\right], \tag{6}$$

where $A(\alpha, \theta)$ depends on both, the hyperangle α and $\theta = \cos^{-1}(\widehat{\mathbf{r}}_2 \cdot \widehat{\mathbf{r}}_3)$, but not on ρ. Thus, the asymptotic hyper-momentum can be thought to be proportional to $1/\sqrt{\rho}$. Solving such a problem requires large spatial domains because the solution itself presents oscillations with very long wavelengths. For the GSF, the challenge lies in the fact that there is not a clearly convenient choice for the basis set energy E_s, since the wave vector varies with the hyperradius ρ. For the present example, the goal was to obtain a solution in a box of 170x170 a.u. using 200x200 Sturmians. We first solved the problem in a preliminary way in a 80x80 a.u. box, with a much smaller basis set (80x80 Sturmians). Then, from the last complete wave inside the preliminary domain, we took the wavelength, and from it, obtained a hyper-momentum at that radius. We finally extrapolated the hyper-momentum from ≈ 70 a.u. to ≈ 170 a.u. following the $\propto \frac{1}{\sqrt{\rho}}$ law, inferring that an energy $E_s = 0.04$ a.u. would be best suited for the bigger calculation.

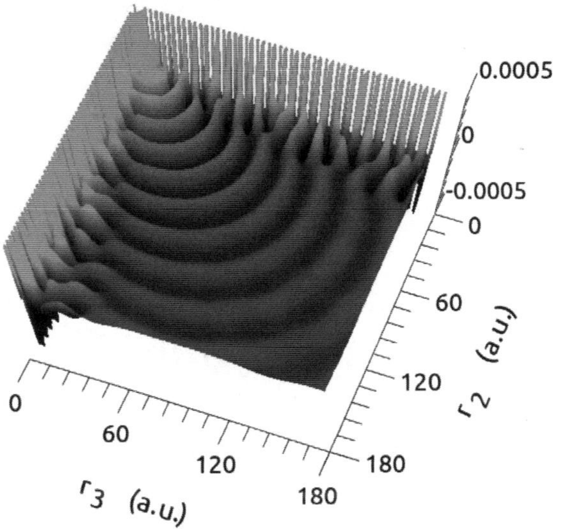

Figure 2. (Color online) Real part of the scattering wave function with zero total energy E_a and momentum transfer $q = 0.24$ a.u.

The real part of the resulting wave function is plotted in figure 2. One can appreciate how the wavelength becomes longer as the hyperradius grows, as indicated by Eq. (6). The slender structures near the two borders correspond to single ionization channels (as was studied in detail in Ref. [10]) and have been truncated in the figure to better visualize the double continuum with zero energy. Double ionization measurements are performed with two electrons emitted with a given energy which can be small (see, e.g., [22, 23]) but finite. Thus, the extreme, more challenging, case $E_a = 0$ serves here only to ensure the GSF method can handle also the very low energy regime for emitted particles.

We have in this section explained how the spherical GSF deals with the double continuum, and illustrated how it could effectively apply to very low emission energies. In the next section we will see the methodology applied to a full three–body case, namely, Eq. (1) beyond the TP approximation.

4. Application to a full three–body continuum problem

Previous work has been done by our group with respect to the long standing open problem (see, e.g., [24]) of the double ionization of Helium by high energy electrons. We first established the theoretical foundations on which we based our scattering solution [9], applied it within a TP framework, and studied the information contained in the wave function [10]. Turning to the full problem, i.e., beyond TP, we first showed results for Single Differential Cross Sections (SDCS) [11] and then for FDCS [25] in the kinematics considered experimentally by Lahmam–Bennani *et al* [12, 13, 14].

Under fast projectile and small momentum transfer kinematics, the transition matrix of the double ionization resembles that of the Double Photo Ionization (DPI) process [13, 14]. Thus, the contribution from the total angular momentum $L = 1$ wave is expected to be the dominant one. Consequently, we describe here a convergence test performed with respect to the $\{l_2, l_3\}$ partial waves corresponding to $L = 1$. The analysis was done in two runs. In the first one, we included partial waves $\{\{0,1\},\{1,2\}\{2,3\},\{3,4\}\}$, and the permuted ones $\{l_2 \leftrightarrow l_3\}$. In the second run, we used $\{\{0,1\},\{1,2\}\{2,3\},\{3,4\},\{4,5\}\}$, and their permutation. In figure 3 we show some of the partial waves contributions to $\Phi^+_{sc}(\mathbf{q},\mathbf{r}_2,\mathbf{r}_3)$, but not the dominant ones for scale purposes. One can see that adding the $\{4,5\}$ pair produced only a very modest modification to the $\{2,3\}$ and $\{3,4\}$ waves calculated in the first run. Moreover, the amplitude of the added $\{4,5\}$ wave is significantly smaller than that corresponding to the pairs $\{0,1\}$ and $\{1,2\}$. Such a contribution would not alter the FDCS in an appreciable fashion, since the amplitudes appear squared, and waves $\{\{2,3\},\{3,4\}\}$ are already a factor ≈ 2 smaller than the dominant ones, $\{\{0,1\},\{1,2\}\}$.

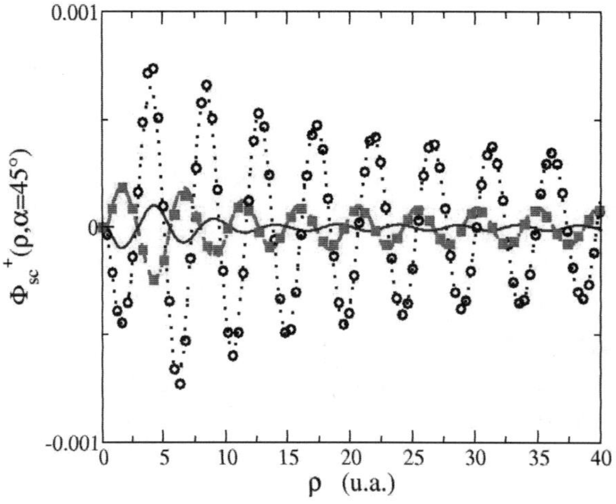

Figure 3. (Color online) Fixed hyperangle $\alpha = \pi/4$ wave function cuts (real part) for $L = 1$; the two ejected electrons possess 10 eV each, and $q = 0.24$ a.u. The wave $\{4,5\}$ (solid line) has a small magnitude when compared to waves $\{2,3\}$ (first run: open circles, second run: dotted line) and $\{3,4\}$ (first run: full squares, second run: dashed line). These waves have in turn roughly half of the amplitude of the more important $\{0,1\}$ and $\{1,2\}$ ones (not shown in the graph).

As introduced in Ref [9], the FDCS can be extracted directly from the asymptotic range of the wave function:

$$\frac{d^5\sigma}{d\Omega_2 d\Omega_3 d\Omega_f dE_2 dE_3} = (2\pi)^3 \frac{k_f k_2 k_3}{k_i \kappa^3} \lim_{\rho\to\infty} \rho^5 |\Phi^+_{sc}(\mathbf{q}, \mathbf{r}_2, \mathbf{r}_3)|^2. \tag{7}$$

From the GSF three-body scattering wave function $\Phi^+_{sc}(\mathbf{q}, \mathbf{r}_2, \mathbf{r}_3)$ we calculated FDCS and present a contour plot in figure 4, for the coplanar Orsay kinematics [14], i.e., $E_2 = E_3 = 10$ eV ejection energies, $E_i = k_i^2/2 = 5599$ eV and $q = 0.24$ a.u.

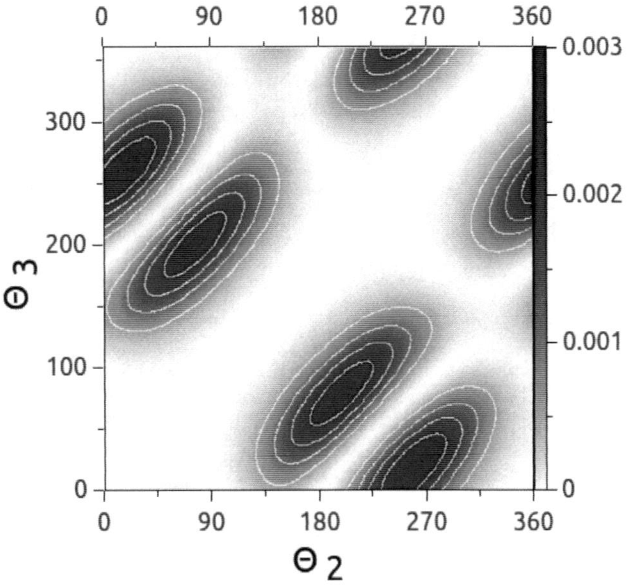

Figure 4. (Color online) Contour plot for the calculated FDCS as a function of both ejection angles θ_2 and θ_3. The kinematics are those used in the Orsay experiment [14] for (10+10) eV emission energies.

We can observe clearly the binary and recoil structures. Because of the small momentum transfer regime, those structures are expected to be present, surrounded by distinct nodal lines detailed in Ref [13]. However, the nonzero momentum transfer q implies that those peaks should not be identical. This is opposed to the DPI case where the momentum transferred to the system by the incident radiation is negligible. figure 4 shows a recoil peak that is indeed more elongated than the binary one. A deeper analysis and comparison with other theories is beyond the scope of this contribution, and can be found elsewhere [25].

5. Concluding remarks

In this contribution we detailed how the spherical GSF basis manages to deal with the three–body continuum by approximating the hyperspherical asymptotic conditions. We have justified and explained how the approach can give accurate results for three–body double continuum physical problems.

We tested the GSF basis with a zero energy problem in a Temkin–Poet model of the electron impact double ionization of Helium. This was done in order to experiment with a demanding problem which is the extreme case of near threshold ionization. The investigation suggests that the GSF method should be able to cope also with low energy, near threshold, kinematics.

In the last section we applied the GSF to the full three–body problem, still in the fast electron impact double ionization of Helium context. The kinematics considered were those of the Orsay experiment [14]. Under the small momentum transfer regime, the double ionization transition matrix has a structure similar to that found in the dipolar DPI case. The $L = 1$ wave is the most dominant one, and as such, we presented a convergence analysis with respect to its $\{l_2, l_3\}$ waves. The FDCS calculated with the GSF method present the expected peak and nodal structure studied by Lahmam–Benanni *et al* [13]. The full analysis of the results in comparison with the experiments and other theoretical calculations are to be presented elsewhere.

Acknowledgments
We acknowledge the CNRS (PICS project No. 06304) and CONICET (project No. Dl 158114) for funding our French-Argentinian collaboration. The support by ANPCyT (PICT08/0934) (Argentina) and PIP 200901/552 CONICET (Argentina) is acknowledged. GG also thanks the support by PGI (24/F059) of the Universidad Nacional del Sur.

References
[1] Frapiccini A L, Gonzalez V Y, Randazzo J M, Colavecchia F D and Gasaneo G 2007 *Int. J. Quantum Chem.* **107** 832–844
[2] Gasaneo G, Ancarani L U, Mitnik D M, Randazzo J M, Frapiccini A L and Colavecchia F D 2013 *Adv. Quantum Chem.* **67** 153–216
[3] Randazzo J M, Buezas F, Frapiccini A L, Colavecchia F D and Gasaneo G 2011 *Phys. Rev. A* **84** 052715
[4] Randazzo J M, Frapiccini A L, Colavecchia F D and Gasaneo G 2009 *Phys. Rev. A* **79** 022507
[5] Randazzo J M, Frapiccini A L, Colavecchia F D and Gasaneo G 2009 *Int. J. Quantum Chem.* **109** 125–134
[6] Frapiccini A L, Randazzo J M, Gasaneo G and Colavecchia F D 2010 *Phys. Rev. A* **82**(4) 042503
[7] Randazzo J M, Ancarani L U, Gasaneo G, Frapiccini A L and Colavecchia F D 2010 *Phys. Rev. A* **81** 042520
[8] Frapiccini A L, Randazzo J M, Gasaneo G and Colavecchia F D 2010 *J. Phys. B* **43** 101001
[9] Gasaneo G, Mitnik D M, Randazzo J M, Ancarani L U and Colavecchia F D 2013 *Phys. Rev. A* **87** 042707
[10] Ambrosio M J, Gasaneo G and Colavecchia F D 2014 *Phys. Rev. A* **89** 012713
[11] Ambrosio M J, Ancarani L, Mitnik D, Colavecchia F and Gasaneo G 2014 *Few-Body Syst.* **55** 825–829
[12] Taouil I, Lahmam-Bennani A, Duguet A and Avaldi L 1998 *Phys. Rev. Lett.* **81** 4600–4603
[13] Lahmam-Bennani A, Taouil I, Duguet A, Lecas M, Avaldi L and Berakdar J 1999 *Phys. Rev. A* **59** 3548–3555
[14] Kheifets A, Bray I, Lahmamm-Bennani, Duguet A and Taouil I 1999 *J. Phys. B.* **32** 5047
[15] Peterkop R K 1977 *Theory of ionization of atoms by electron impact* (Colorado Associated University Press)
[16] Kadyrov A S, Mukhamedzhanov A M, Stelbovics A T, Bray I and Pirlepesov F 2003 *Phys. Rev. A* **68** 022703
[17] Kadyrov A S, Mukhamedzhanov A M, Stelbovics A T and Bray I 2004 *Phys. Rev. A* **70** 062703
[18] Gasaneo G and Ancarani L U 2012 *J. Phys. A* **45** 045304
[19] Mitnik D M, Gasaneo G and Ancarani L U 2013 *J. Phys. B* **46** 015202
[20] Mitnik D M, Gasaneo G, Ancarani L U and Ambrosio M J 2014 *J. Phys. : Conf. Ser.* **488** 012049
[21] Klar H and Schlecht W 1976 *J. Phys. B* **9** 1699
[22] Dürr M, Dorn A, Ullrich J, Cao S P, Czasch A, Kheifets A S, Götz J R and Briggs J S 2007 *Phys. Rev. Lett.* **98**(19) 193201
[23] Ren X, Dorn A and Ullrich J 2008 *Phys. Rev. Lett.* **101** 093201
[24] Ancarani L U, Cappello C D and Gasaneo G 2010 *J. Phys: Conf. Ser.* **212** 012025
[25] Ambrosio M J, Colavecchia F D, Gasaneo G, Ancarani L U and Mitnik D M 2014, submitted to *J. Phys. B*

Electron impact experimental and theoretical fourfold differential cross section for CH_4

R El Mir[1,2,3], E M Staicu Casagrande[3,4], C Dal Cappello[1], A Naja[2], A Mansouri[5] and I Kada[5]
[1] SRSMC, UMR CNRS 7565, Université de Lorraine, BP 70239, F-54506 Vandoeuvre-les Nancy, France
[2] LPM, EDST, Lebanese University, Tripoli, Lebanon
[3] Université Paris-Sud, Institut des Sciences Moléculaires d'Orsay (ISMO), Bât. 351, F-91405 Orsay Cedex, France
[4] CNRS, ISMO, Bat. 351, F-91405 Orsay Cedex, France
[5] LPQSD, Département de Physique, Faculté des Sciences, Université de Setif1, 19000, Algeria

E-mail: elena-magdalena.staicu-casagrande@u-psud.fr

Abstract. New development of the 3C approximated model describing the fourfold differential cross section for double ionization of CH_4 will be reported. A corrective factor called Ward-Macek will take into account the repulsion between the two ejected electrons instead of the usual Gamow factor. To validate the developed model, we performed the coplanar (e, 3-1e) experiments for the double ionization of the methane target by electron impact at intermediate incident energy and different energy range for the pair of ejected electrons. Some agreement is found for the forward part of the angular distributions. The differences are discussed in terms of double ionization mechanisms.

1. Introduction

Collision has always been the main process leading to several discoveries in quantum world. The study of double ionization (DI) responds to the ever-increasing need for reliable cross-section data in diverse fields not only in physics but also in life sciences. DI by electron impact, called (e,3e) experiments see e.g. [1-6], is a main example of collision process during which all the electrons issued are determined in *triple coincidence* (or alternatively the residual ion plus two electrons [1,2]). Measuring the fully differential cross section for DI by performing kinematically complete experiments is a big challenge because of its small value. It is much easier to perform (e, 3-1e) experiments where only two of the three electrons in the final state are detected in *double coincidence* using the corresponding ionization potential IP^{2+} of the DI process.

Measurements of fourfold differential cross section (4DCS) for DI of rare gases such as He, Ne and Ar were reported in previous studies [1,7,8]. The corresponding results were obtained for coplanar geometry with equal and unequal energy sharing of ejected electron. Despite the difficulties imposed by the molecular nature, we published measurements of single ionization (SI) process of CH_4 [9] and compared the results to first order theoretical approaches. In addition, experiments of DI on methane have been performed with equal and unequal ejected electron energies (E_b : E_c) = (37 : 37) eV and (E_b : E_c) = (37 : 12) eV respectively [10]. Experimental results were compared to an approximation of the well-known 3C model, labelled 2CWG, where the Gamow factor was used to describe the repulsion between the two ejected electrons. At the present time, only first order models are available because of the complexity nature of the molecular target.

The aim of the present work is to develop the theoretical approach used, leading to a better analysing of the molecular influence on the DI process. Note that the Gamow factor which is used generally in the 2CWG model, underestimates the 5DCS's amplitude especially when the two ejected electrons have the same ejected energy [11]. For this purpose we replaced it by a corrective one, obtaining the 2CWWM model where the Ward-Macek factor is used instead of the Gamow factor. The second goal of this work is to validate our theoretical model, by comparing the theory with the (e, 3-1e) experimental data already published but also with new experiments with equal energy sharing of ejected electrons ($E_b : E_c$)=(12 : 12) eV.

2. Experiments

The experimental set-up used for the present experiments has been extensively described in [12] and it is identical to those used in [7,8]. Briefly, a monochromatic incident electron beam crosses the gas jet; the scattered electron is then detected in coincidence with the ejected electrons emitted in the collision plane defined by the incident and scattered momentum vectors, $\vec{k}_0$ and $\vec{k}_a$ respectively. This arrangement is known by 'coplanar geometry'. The electron-electron coincidence spectrometer is mainly characterized by three multi-angle toroidal analyzers coupled with positive sensitive detectors. Throughout this work, positive angles are counted clock wise starting from the incident electron beam direction. The 'fast' scattered electron, labelled 'a', is detected with a fixed energy E_a=500 eV at two symmetrical angles θ_a= + (6° ± 3°) and θ_a= - (6° ± 3°). Among the two ejected electrons resulting from the double ionization DI of the target (labelled 'b' for the faster and 'c' for the slower), we choose to detect only the faster with energy E_b in coincidence with the a-scattered electron, hence an (e, 3-1e) experiment. Of course, such distinction does not hold in the case of equal ejection energies, but we keep labelling 'b' the detected electron. The b-ejected electron is detected in almost the full angular range of the collision plane, i.e. θ_b= 20°- 160° and θ_c= 200°- 340° where 0° is defined by the incident beam direction. Although the emission direction of the third c-electron is unknown, its kinetic energy E_c is known from the energy conservation $E_0 - IP^{2+} = E_a + E_b + E_c$. The energy resolution of each analyser is set to 3 eV.

The experiments for CH_4 ($1t_2$) were performed with equal ejected electron energies ($E_b : E_c$) = (37 : 37) eV and (12 : 12) eV and unequal energies ($E_b : E_c$) = (37 : 12) eV. The experimental parameters are listed in table 1. The incident energy E_0 is consequently adjusted to fulfill the energy conservation requirement for the methane target under study, with a DI potential of the outermost orbital $1t_2$ IP^{2+} equal to 38eV. Here it is assumed that the DI process removing two electrons from the outermost orbital and leaving the ion in its ground state is dominant, but DI of the other orbital such as 2a1 cannot be excluded.

Table 1. Kinematical parameters for the DI process of CH_4 considered in this study. The last line indicates the forward and backward directions of ejection of the 'b'-electron (detected electron).

	E_a = 500 eV	θ_a=-6°	
E_0 (eV)	612	587	562
E_b (eV)	37	37	12
E_c (eV)	37	12	12
K (a.u)	0.93	0.83	0.75
θ_K / θ_{-K} (deg)	43/223	50/230	58/238

3. Theoretical framework

In the first Born approximation the six-fold differential cross section (6DCS) is given by

$$\sigma^6(\alpha;\beta;\gamma) = \frac{d^6\sigma(\alpha;\beta;\gamma)}{d\Omega_{Euler}d\Omega_a d\Omega_b d\Omega_c dE_b dE_c} = \frac{k_a k_b k_c}{k_0}|f_{B1}|^2 \quad (1)$$

for a molecular orientation defined by the Euler angles $(\alpha;\beta;\gamma)$ [12] where $d\Omega_{Euler} = sin\beta\, d\beta\, d\alpha\, d\gamma$ and $d\Omega_a$, $d\Omega_b$ and $d\Omega_c$ denote the elements of solid angles for the scattered and the two ejected electrons b and c, respectively, whereas the energy intervals of the ejected electrons are represented by dE_b and dE_c. The momenta of the incident, the scattered, and the two ejected electrons are denoted by $\vec{k}_0$, $\vec{k}_a$, $\vec{k}_b$ and $\vec{k}_c$ respectively. In a (*e*,3*e*) reaction the conservation of energy imposes: $\frac{k_0^2}{2} = \frac{k_a^2}{2} + \frac{k_b^2}{2} + \frac{k_c^2}{2} + IP^{2+}$. The first Born term f_{B1} is written as

$$f_{B1} = -\frac{1}{2\pi}\langle \exp(i\vec{k}_a.\vec{r}_0)\,\Psi_f(\vec{k}_b,\vec{k}_c,\vec{r}_1 \dots \vec{r}_{10})|V|\exp(i\vec{k}_0.\vec{r}_0)\,\Phi_i(\vec{r}_1 \dots \vec{r}_{10})\rangle \quad (2)$$

where $\Phi_i(\vec{r}_1 \dots \vec{r}_{10})$ is the wave function of the initial state of the methane molecule and $\Psi_f(\vec{k}_b,\vec{k}_c,\vec{r}_1 \dots \vec{r}_{10})$ is the wave function for the double continuum state of the methane molecule. We here neglect the exchange effects between the incident particle and the ejected electrons because both the incident and the scattered electrons are faster than any ejected one. In equation (2), the potential V represents the Coulomb interaction between the incoming electron and the target and is written as

$$V = -\frac{6}{r_0} - \frac{1}{|\vec{r}_0 - \vec{R}_1|} - \frac{1}{|\vec{r}_0 - \vec{R}_2|} - \frac{1}{|\vec{r}_0 - \vec{R}_3|} - \frac{1}{|\vec{r}_0 - \vec{R}_4|} + \sum_{i=2}^{10}\frac{1}{r_{0i}} \quad (3)$$

where $R_1 = R_2 = R_3 = R_4 = R_{CH}$, $\vec{r_i}$ denotes the position vector of the ith bound electron of the target with respect to the centre of the carbon nucleus, $\vec{r_0}$ the coordinate of the incident particle and $\vec{r_{0i}} = \vec{r_0} - \vec{r_i}$.

The reduction of this 10-electron target problem to a two-electron target may be done by using the frozen-core approximation. In this case, the two target electrons are those which will be ejected during the double ionization. Then, the initial wave function describing the two active electrons is $|\Phi_i(\vec{r}_1,\vec{r}_2)\rangle$ and is given by the single centre wave function of Moccia [14]. In the present work, we are considering a DI process of the three orbitals ($1t_{2x}$, $1t_{2y}$ and $1t_{2z}$) of methane molecule (vapour gas). The 15 possible ion final states of CH_4^{++} have been considered. The final state wave function, which describes the two ejected electrons, is an approximate BBK wave function (called 2CWWM model) and it is given by:

$$\Psi_f(\vec{k}_b,\vec{r}_1,\vec{k}_c,\vec{r}_2) = \frac{1}{\sqrt{2}}\left[\Psi_c^-(\vec{k}_b,\vec{r}_1)\Psi_c^-(\vec{k}_c,\vec{r}_2) \pm \Psi_c^-(\vec{k}_b,\vec{r}_2)\Psi_c^-(\vec{k}_c,\vec{r}_1)\right] \times \varphi(|\vec{k}_b - \vec{k}_c|) \quad (4)$$

The sign + corresponds to a wave function where the spin part is antisymmetric, while the sign – refers to a symmetric spin part. Note that

$$\Psi_c^-(\vec{k}_b,\vec{r}_1) = \frac{1}{(2\pi)^{\frac{3}{2}}}\exp(i\vec{k}_b.\vec{r}_1)\,\Gamma(1-i\alpha)\exp\left(-\frac{\pi}{2}\alpha\right)\;{}_1F_1(i\alpha,1,-i(\vec{k}_b.\vec{r}_1 + k_b r_1)) \quad (5)$$

where $\alpha = -Z/k_e$ and Z=2 while the Ward-Macek factor [15] presented in Eq.(4), is

$$\varphi(|\vec{k}_b - \vec{k}_c|) = \exp(-\frac{\pi\chi_{bc}}{2})\Gamma(1-i\chi_{bc})\;{}_1F_1(i\chi_{bc},1,-i|\vec{k}_b - \vec{k}_c|r_{12av})$$

with $\chi_{bc} = \frac{1}{|\vec{k}_b - \vec{k}_c|}$ and r_{12av} is an average of the distance $\vec{r}_{12} = \vec{r}_1 - \vec{r}_2$.

The integration over the projectile coordinates (r_0) can be performed analytically. In order to make a comparison with (*e*,*3e*) experiments we must average the above-described 6DCS over the Euler angles in order to obtain the total cross section for DI (5DCS), namely

$$\sigma^{(5)} = \frac{d^5\sigma}{d\Omega_a d\Omega_b d\Omega_c dE_a dE_b} = \frac{1}{8\pi^2}\iiint d\alpha\, d\gamma\, d\beta\, sin\beta\,[\sigma^6(\alpha;\beta;\gamma)] \quad (6)$$

Knowing that DI proceeds by different pathway, it is common to consider that 5DCS is mainly dominated by three different mechanisms: shake-off, two-step 1 and two-step 2 known by SO, TS1 and

TS2 respectively. The two first-order mechanism SO and TS1 are included in the first Born approximation in a partial way [16]. The second-order process in the projectile-target interaction (TS2) can only be described by the second Born approximation B2 which needs a difficult eight numerical integration. Finally, four-fold differential cross section (4DCS) calculations defined as

$$\sigma^{(4)} = \iint \sigma^5 d\Omega_c = \frac{d^4\sigma}{d\Omega_a d\Omega_b dE_a dE_b} \tag{7}$$

were performed. It corresponds to the coincident detection of the scattered electron and only the 'faster' ejected electron.

4. Results and discussion

Experimental and theoretical results of fourfold differential cross section for DI (e, 3-1e) of the outermost orbital $1t_2$ of CH_4 at $(E_b : E_c) = (37 : 37)$ and $(12 : 12)$ are shown in figures 1 (a)-(b) respectively and 4DCS of CH_4 at $(E_b : E_c) = (37 : 12)$ eV is shown in figure 2.

Relative experimental data have been normalized to the same absolute value of theory at the maximum of the forward lobe. Identically to the observation made in our previous work [9], we note some important common features: (i) the displaying of the angular distribution of the 4DCS of two-lobe structure: a forward lobe pointing roughly in the momentum transfer direction (θ_K) and a backward lobe pointing in the opposite direction (θ_{-K}), (ii) the breaking of symmetry about $\pm K$ and (iii) the exhibition of additional structures in these lobes.

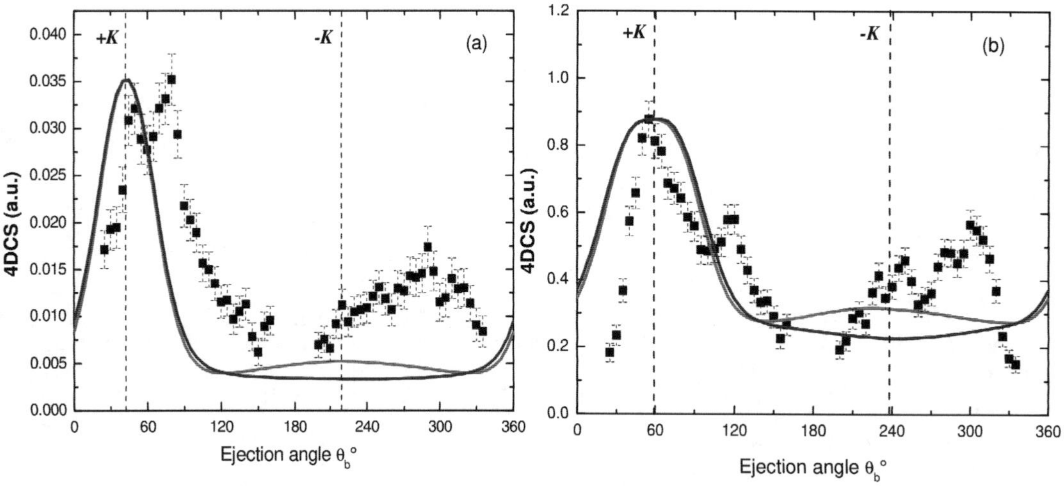

Figure 1. Four-fold differential cross sections (4DCS) for the double ionization of CH_4. The scattered electron with energy E_a = 500 eV is detected at an angle θ_a = -6° in coincidence with the fast-emitted electron with energy E_b, whereas the slow-emitted electron with energy E_c remains undetected. Panel (a): $(E_b : E_c) = (37 : 37)$ eV, (b): $(E_b : E_c) = (12 : 12)$ eV. Full squares are the experimental data, with one standard deviation statistical error bar. Blue and red solid curve present theoretical predictions from the first-order Born model using Ward-Macek factor and Gamow factor respectively. The absolute scale shown is that of 2CWWM to which the experiment is normalized for best visual fit at the maximum of the forward lobe. Results obtained by 2CWG were multiplied by an arbitrary factor to fit the maximum of the forward lobe of 2CWWM. This factor is equal to 6.4 and 18.1 respectively for the cases $(E_b : E_c) = (37 : 37)$ eV and $(12 : 12)$ eV. The thin dashed lines indicate the direction of the momentum transfer (θ_K) and its opposite.

The two sets of the experimental data with equal energy sharing for the ejected electrons, $(E_b : E_c) = (37 : 37)$ eV and $(E_b : E_c) = (12 : 12)$ eV, show some differences in the intensity of the backward lobe with respect to the forward one: when the energy of the outgoing electrons is lowered, the ratio between forward and backward lobe is decreasing. Another difference seems to be related to the width of the forward lobes: in the (12 : 12) case, the forward lobe is larger than in the (37 : 37) case. The post collision interactions which can explain this behavior are included in the theoretical model. We observed indeed that the theoretical results follow the same trend like the experimental results. However, in the comparison with the theoretical results important differences are found between this first-order theory and experiments:

- the existence of the backward structures which are not predicted by the B1 calculations;
- small additional structures in the forward lobe;
- in the (37 : 37) eV case an important shift with respect to the $\pm K$ directions of the two main lobes. In the (37 : 12) and (12 : 12) eV cases, this shift is less pronounced for the forward distribution, probably due to the fact that this lobe is much larger than in the (37 : 37) eV case and probably includes more contribution of the B1 mechanisms.

In the 4DCS of CH_4 at $(E_b : E_c) = (37 : 12)$ eV important differences between the experiment and the theory are also presented. Experimental results show a second structure in the forward lobe and the backward lobe even if it is less important than in the other cases. This second structure is not reproduced by the theory and may be associated to the contribution of the high-order mechanisms.

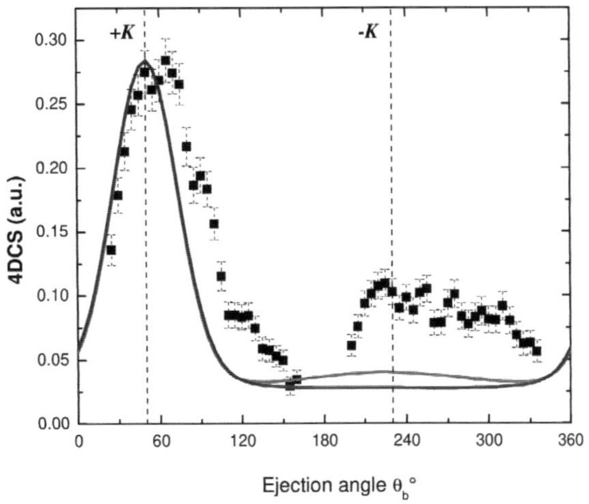

Figure2. Four-fold differential cross sections for (e, 3-1e) double ionization of CH_4 for unequal energy sharing $(E_b : E_c)$ =(37 : 12) eV. Full squares are the experimental data, with one standard deviation statistical error bar. Blue and red solid curve present theoretical predictions from the first-order Born model using Ward-Macek factor and Gamow factor respectively. The absolute scale shown is that of 2CWWM to which the experiment is normalized for best visual fit at the maximum of the forward lobe. Results obtained by 2CWG were multiplied by an arbitrary factor to fit the maximum of the forward lobe of 2CWWM. In the present case, this factor is equal to 9.6. The thin dashed lines indicate the direction of the momentum transfer (θ_k) and its opposite.

To date, no theoretical second-order calculations for the present CH_4 (e, 3-1e) measurements are available.

5. Conclusion

In this paper we present the (e, 3-1e) studies for double ionization on methane. One of the motivations of this study is to present the last theoretical results obtained with a new way to describe the repulsion between the two ejected electrons. The experimental data have been compared with a B1 theoretical model which can reproduce the forward lobe of the angular distributions. The structures related to the

Journal of Physics: Conference Series **601** (2015) 012005 doi:10.1088/1742-6596/601/1/012005

B1 mechanism are well described by the theory but additional structures appear in both forward and backward lobes. This suggests that other mechanisms are involved in the ionization and there are revealing limitations in the theoretical model which doesn't take into account the TS2 mechanism for instance.

References

[1] Lahmam-Bennani A, Staicu Casagrande E M, Naja A, Dal Cappello C and Bolognesi P 2010 *J. Phys. B: At. Mol. Opt. Phys.* **43** 105201
[2] Lahmam-Bennani A, Dupre C and Duguet A 1989 *Phys. Rev.Lett.* **63** 1582
[3] Dupre C, Lahmam-Bennani A and Duguet A 1991 *Meas. Sci.Technol.* **2** 327
[4] Taouil I, Lahmam-Bennani A, Duguet A and Avaldi L 1998 *Phys. Rev. Lett.* **81** 4600
[5] Jia C C, Lahmam-Bennani A, Duguet A, Lecas M and Dal Cappello C 2002 *J. Phys. B: At. Mol. Opt. Phys.* **35** 110313
[6] Naja A, Staicu Casagrande E M, Ren X G, Catoire F, Lahmam-Bennani A, Dal Cappello C and Whelan C T 2007 *J. of Phys. B: At. Mol. Opt. Phys.* **40**, 2871-84
[7] Staicu Casagrande E M, Li C, Lahmam-Bennani A, Dal Cappello C, Schultz M and Ciappina M 2011 *J. Phys. B: At. Mol. Opt. Phys.* **44** 055201
[8] Li C, Lahmam-Bennani A, Staicu Casagrande E M and Dal Cappello C 2011 *J. Phys. B: At. Mol. Opt. Phys* **44** 115201
[9] Nixon K L, Murray A J, Chaluvadi H, Amami S, Madison D H and Ning C 2012 *J. Chem. Phys.* **136** 094302
[10] Staicu Casagrande E M, Li C, Lahmam-Bennani A and Dal Cappello C 2014 *J. Phys. B: At. Mol. Opt. Phys.* **47** 115203
[11] Dal Cappello C and Ancarani L U 2008 *J. Phys. B: At. Mol. Opt. Phys.* **41** 208001
[12] Catoire F, Staicu Casagrande E M, Lahmam-Bennani A, Duguet A, Naja A, Ren X G, Lohmann B and Avaldi L 2007 *Rev. Sci. Instrum.* **78** 013108
[13] Kada I, Mansouri A, Dal Cappello C, Hervieux P A and Roy A C 2009 *J. Phys. B: At. Mol. Opt. Phys* **42** 025201
[14] Moccia R 1964 *J. Chem. Phys. A.* **40** 2186
[15] Ward S J and Macek J H 1994 *Phys. Rev. A* **49** 1049
[16] Popov Yu V, Dal Cappello C, Joulakian B and Kuzmina N M 1994 *J. Phys. B: At. Mol. Opt. Phys* **27** 1599

Investigating dynamics of complex system irradiated by intense x-ray free electron laser pulses

L Fang[1,2], Z Jurek[3,4], T Osipov[1], B F Murphy[1], R Santra[3,4,5] and N Berrah[6]
[1] Physics Department, Western Michigan University, Kalamazoo, MI 49008 USA
[2] The Center for High Evergy Density Science, Unvierstiy of Texas at Austin, Austin TX 78712 USA
[3] Center for Free-Electron Laser Science, DESY, 22607 Hamburg, Germany
[4] The Hamburg Centre for Ultrafast Imaging, 22761 Hamburg, Germany
[5] Department of Physics, University of Hamburg, 20355 Hamburg, Germany
[6] Physics Department, University of Connecticut, Storrs, CT 06269 USA

E-mail: Nora.berrah@uconn.edu

Abstract. We carried out experimental and theoretical investigation of the response of a complex molecule, C_{60}, to intense x-ray photon beam from a free-electron-laser. We show good agreement between the modelling and the experiment. Our model, which can be scaled well to larger systems, reveals femotosecond molecular dynamics details, at the level of atomic resolution, which are inaccessible directly by our experiments. Our results illustrate the variety of physical and chemical processes in the interaction between large molecules and intense x-ray pulses, including photoelectric effect, secondary ionization, recombination and inter-atomic Auger decays. The understanding of these processes has a broad impact on research that implements intense x-ray pulses.

1. Introduction

The recent development of intense photon sources in the x-ray regime and femtosecond time scale, allows the understanding of the fundamental response of atoms and molecules to these photon beams underpining a broad range of research, that includes biology and chemical sciences [1]. While the investigations of the behavior of atoms and small molecules exposed to the extreme condition set by the intensity of free electron lasers (FELs) have just been launched, the exploration of larger molecules is even more nascent due to the complexity of the molecular structures and the interaction processes.

We carried out experimental and theoretical investigation on the photon interaction of a midsize molecule, C_{60}, using x-ray pulses from the Linac Coherenet Light Source (LCLS) FEL. The experiment was performed at various photon beam parameters. Our theoretical model was validated based on the good agreement between the experimental and the simulation results in terms of molecular and atomic ion yields and fragment atomic kinetic energies. Our model presents various dynamical molecular ionization processes that are absent in atomic scenario. It reveals with atomic resolutions, the evolution of the molecular electronic and nuclear structure, i.e., the explosion of the molecule, and the evolution of the ionization stages and internal potentials during and after the photon-interaction. We show that the contribution of several effects, such as secondary ionization, recombination and molecular chemical bonds, are very important in comparison to atomic scenario and modeling with inclusion of selected molecular effects [2].

The dominant ionization mechanism for atoms and molecules irradiated by intense x-ray photon beams is sequential ionization - multiple sequential core-level ionization followed by subsequent Auger decay, which has been observed in atoms [3] as well as in small molecules [4]. For a large system such as C_{60}, many electrons are generated via photoionization, Auger decays and secondary ionization. Some of them leave the system, leading to a highly charged positive ion that may trap other ejected electrons. Chemical bonds of various strength between carbon atoms also play a role in determining the ionization and fragmentation pathways, adding complexity to the photon-interaction dynamics and hence, adding difficulties in predicting the final outcomes. All these processes which occur at femtosecond time scales contribute to the experimental observation, but, in the current experiment, the detailed dynamical information is lost, because we used a single pulse and the pulse duration is comparable to the time scale of the dynamical processes themselves. However, with a model validated by direct comparison with experimental quantities, we are able to take snapshots of the dynamics and therefore, gain insights into the evolution of the system, i.e. what the contribution of the various processes is.

2. Experiment and theoretical model

The experiment was performed at the atomic, molecular and optical physics (AMO) hutch of the Linac Coherent Light Source (LCLS) at SLAC National Accelerator Laboratory using the High Field Physics instrument [5]. The x-ray beam was focused to achieve a peak focal intensity of 10^{16}–10^{18} W/cm^2 and crossed a collimated molecular beam of C_{60} molecules from a resistively heated oven at the focus. A magnetic bottle spectrometer was used for kinetic energy (KE) resolved ion time-of-flight spectroscopy with high collection efficiency, even at several hundred eV ion KE.

In the last decade, several theoretical models have been developed to study the evolution of samples irradiated by x-ray FEL pulses [6]. For simulation of large systems under these extreme conditions which are likely to end up in highly excited states, fully quantum mechanical methods are numerically not feasible today. In the current work, we developed a molecular dynamics (MD) model, where we utilized the XMDYN [7] tool to model finite samples irradiated by high intensity x-ray pulses. In our model, we used an atomistic description of C_{60} combined with a molecular dynamics treatment of the real space dynamics. The electronic configurations of the individual atoms/ions were tracked and utilized. Cross sections and rates for photoionization, fluorescent and Auger relaxation processes were calculated by the XATOM toolkit [7] and the Monte Carlo algorithm was implemented to describe these stochastic events. Within the model, electrons and ions are treated using classical mechanics. The fullerene-specific classical Brenner force field accounted for the chemical bonds between atoms, and the Coulomb forces for the interaction between the charged particles. The real-space evolution of the system was obtained by numerically solving the Newton equations.

For a direct comparison of experiment and theory, the actual x-ray FEL pulse properties at the interaction region must be known. Calibrations revealed that the x-ray FEL parameters can differ from the nominal values provided by LCLS, e.g. the actual pulse duration is 50–60% of the nominal value as also indicated by earlier work [3]. Our calibrations indicated that for the nominal pulse durations of 7, 20, 60, and 150 fs full width at half maximum recorded with the experimental data, actual pulse durations of 4, 13, 30, and 90 fs respectively are more appropriate. We use throughout this paper the 'actual' calibrated pulse durations, and we refer to these as short (4 fs), intermediate (13, 30 fs), and long pulses (90 fs).

3. Results and discussion

3.1. Comparison between experimental results and simulations – validation of a model for understanding the dynamics of large systems under extreme conditions

We measured both molecular ions of various charges and atomic ions from 1+ to up to 6+ as the outcome of the ionization by the LCLS x-ray pulses at 485 eV and 600 eV. The atomic C ions are part of the final products subsequent to the various molecular processes the effect of which was imprinted in the ion yields and KE. The FEL beam has a spatially non-homogeneous intensity distribution. According to modelling full atomic fragmentation happens at the highest intensities, while molecular fragments originate from lower fluence regions. Therefore the atomic species are our main focus in the comparison between the modeling and the experiment. We note, however, that low charge state atomic ions, in particular C^{1+}, are also formed at lower fluences. The atomic charge state distributions change when varying the x-ray beam parameters, such as the pulse energy and the pulse duration. Figure 1

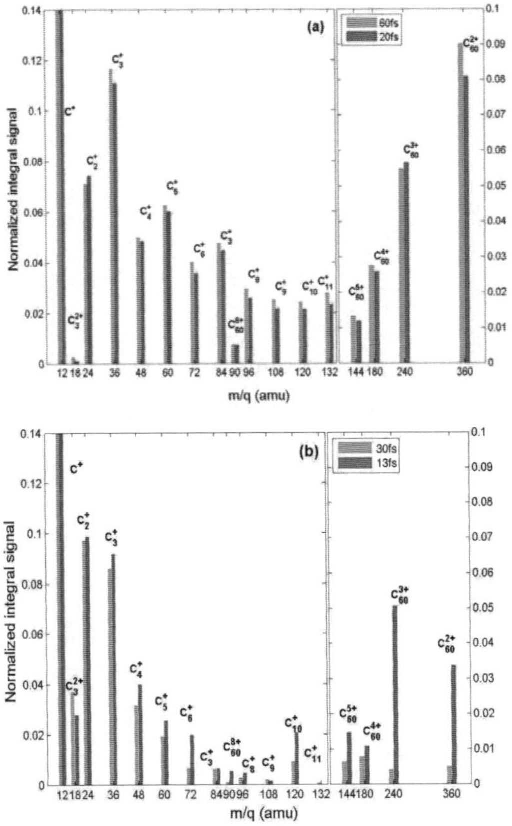

Figure 1. (a) Experimental integral signals of molecular peaks at two pulse duration. The photon energy is 600 eV; the pulse energy is 0.61mJ. The C^+ peak is out of scale in the current plots. (b) Calculations using parameters that model best the measurements at intermediate pulse duration. The figure is adapted from [8].

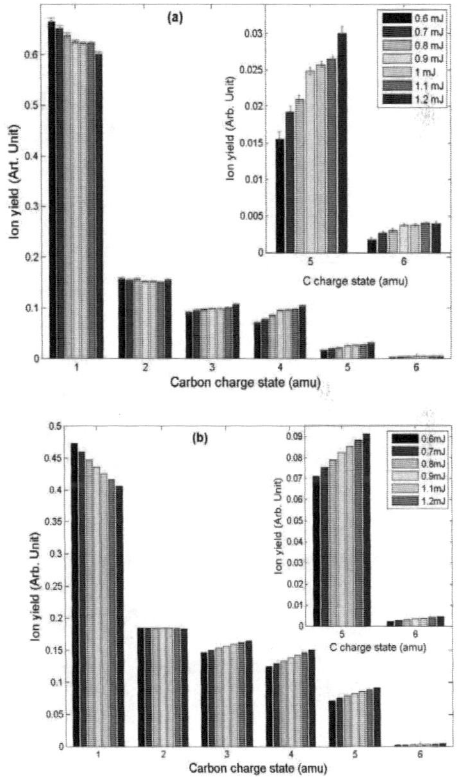

Figure 2. (a) Experimental charge state distribution of atomic C ion fragments at various pulse energies and the same (long) pulse duration. The photon energy is 485 eV. (b) Calculated charge state distribution. Figure is adapted from [8]. Experimental data were obtained by direct summation of the signals by x-ray shots of particular pulse energies as shown in the legend. For each charge state, bars from left to right correspond to energies from low to high.

shows experimental (upper panel) and simulated (bottom panel) yields of various molecular ions (and atomic C^+ ion) at two different pulse durations. Though there are similar trends in the experimental and theoretical data, the appearing discrepancy indicates that there is still challenge for theory to capture fragmentation dynamics in this region. On the other hand our model successfully reproduces the observed atomic charge state distribution at various x-ray beam parameters, as seen in figure 2. Figure 2 shows experimental results (upper panel) and simulation results (bottom panel) of atomic carbon ion yields for all charge states at different pulse energies. The simulation of the pulse energy dependence of the ion yields for various charge states matches qualitatively the experimental data.

We also compare the atomic ion KE from simulations to the experimental data. Figure 3 shows the comparison between the model and experimental KE of atomic C ions generated at the high fluence part of the x-ray pulse (except that C^+ was produced at low fluence). Measured ion KE distributions show good agreement with the calculations in both mean ion energy and RMS energy spread for short and intermediate pulse durations at near constant x-ray fluence. Agreements between experiments and the simulation in various aspects validate our MD model.

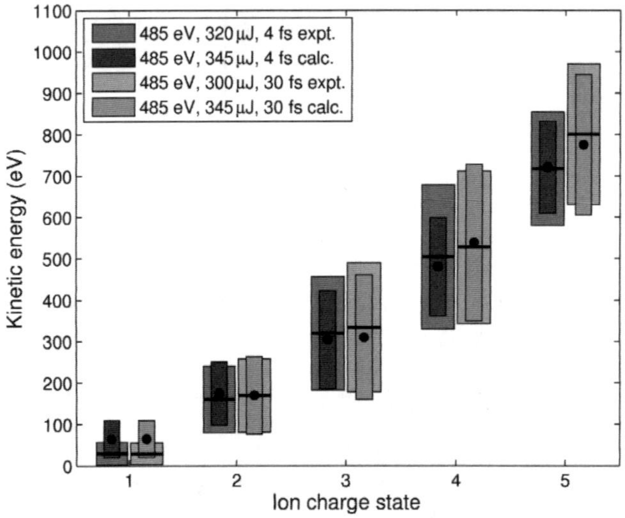

Figure 3. Fragment ion kinetic energy versus charge state. Experimental data (wide rectangles) and calculations (narrow rectangles) are displayed for short and intermediate duration pulses. Mean ion kinetic energy versus charge state is indicated by lines (experiment) and circles (simulation) at the center of each rectangle, while RMS kinetic energy width is indicated by the height of each rectangle. Pulse energy in simulation is 345 μJ compared to experimental data taken at 320 μJ (short pulses) and 300 μJ (intermediate pulses). The figure is adapted from [2].

3.2. Evolution of expanding C_{60} moleucule irridiated by intense x-ray pulses

With our MD model, we simulated the electronic and nuclear dynamic evolution of expanding C_{60} molecules irradiated by LCLS, as shown in figure 4. Early in the x-ray laser pulse, due to photoionization and Auger electron emission, high energy electrons appear that can escape, leaving behind a highly charged C_{60} ion accumulating Coulomb potential energy (figure 1(a)). C_{60} starts to explode due to the increasing ion repulsion, and substantial atomic displacement (~10 Å) occurs within the first tens of femtoseconds (figure 4(b)(c)). As the ionic potential increases, later on during the pulse, some of the photo- and Auger electrons, together with slow electrons generated via secondary ionizations, are trapped by the increasing ionic potential, forming a nanoplasma surrounding and penetrating the remnant of the buckyball, while electrons that could escape are already at large distances (figure 4(c)). The electron nanoplasma expands with the carbon ions and a fraction of it evaporates within nanoseconds, while others recombine forming the final detectable ionic charge states (figure 4(d)(e)).

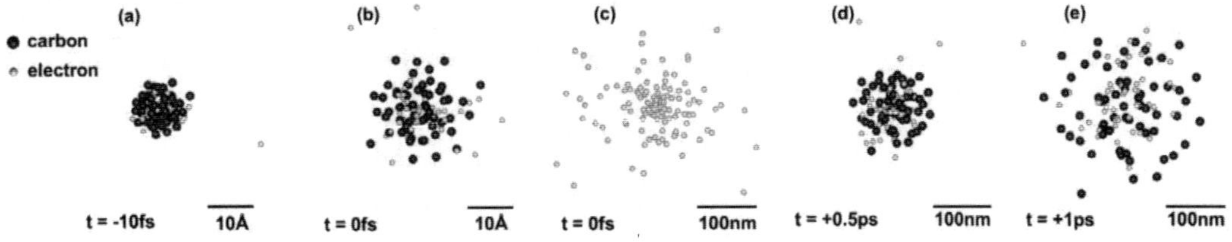

Figure 4. Simulated real space snapshots of the time evolution of a C_{60} molecule irradiated at the center of the focus of the x-ray free electron laser pulse. Spatial distribution of carbon atoms (solid dark colar spheres) and electrons (solid light color spheres) are shown as a function of time at fine spatial scale (a, b) and expanded spatial scale (c, d, e). The centre of the pulse is at t = 0 fs and the photon energy is 600 eV with 30 fs pulse duration and 0.6 mJ pulse energy. The figure is adapted from [8].

We tracked the time evolution of the average radius of C_{60} and of the charge per C atom. Figure 5 shows the average radius of the system and the KE per atom as a function of time. At all pulse fluences used in the simulations, within the first 100 femtoseconds, the fragment ions reach their asymptotic velocity having converted most of the accumulated potential energy into kinetic energy. At higher fluences, it takes less time to reach the asymptotic velocity.

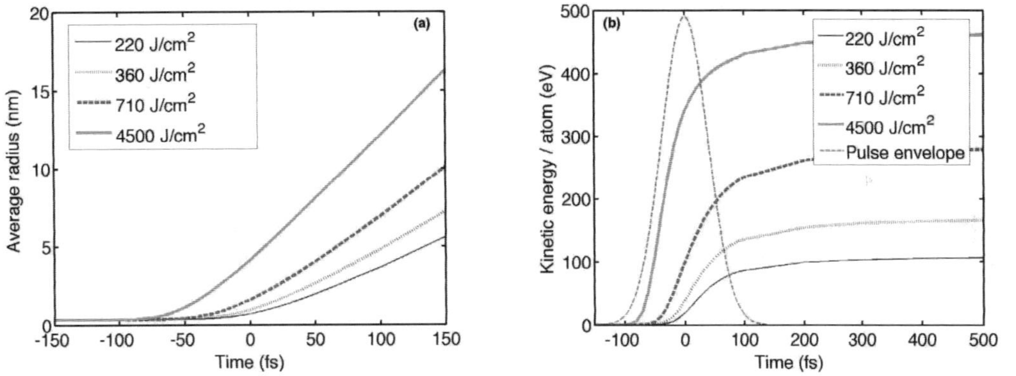

Figure 5. Calculated average molecular radius (a) and KE per atom (b) as a function of time at different pulse fluences. The photon energy used is 485 eV and the pulse duration used is 90 fs (long pulses). The molecules expand from an initial average radius of 355 pm. Fluence above 360 J cm^{-2} produces full atomic fragmentation in our calculations. Time zero corresponds to the center of the pulse. The figure is adapted from [2, 8].

3.3. Atomic vs molecular scenario – demonstrating molecular ionization mechanisms

To show the molecular effects that influence the properties of the atomic ions resulting from C_{60} ionization and fragmentation, we compare the theoretical results regarding atomic charge states using the ionization of atomic C and the ionization of C_{60}. Figure 6 shows the atomic ion yields obtained from the ionization of atomic C and for molecular C_{60} using the MD model. With the atomic model, the ion yields present an alternation pattern in the intensities between odd and even charges states till the Auger decay is not possible due to lack of valence electron at charge larger than 3. This is seen in experiments with atoms [3] where interatomic Auger decay is absent and is resultant from the photoionization and Auger event sequence that removes two electrons at a time. For molecules, this pattern breaks down, since molecular ions can fragment following various fragmentation pathways and interatomic Auger decay can occur as well. For example, dissociation of a C_2^{2+} ion leads to 2 C^+ ions. According to the results of the molecular model, we see a monotonic decrease of ion yields for higher charge stages with the highest abundance in C^+, in huge contrast with the result from the atomic model.

As mentioned above, the discrepancy between the results from the atomic and molecular models is partially due to electrons' redistribution during dissociation and ending up in different atoms. There are other mechanisms that are absent in atomic ionization that contribute to the final outcomes. Beside the photoionization and Auger relaxation, secondary ionization and electron recombination events take part in changing the charge states, as indicated in figure 7. We have calculated the number of ionization events associated with various ionization mechanisms using an atomic and a molecular model for long and short pulse durations (figure 7). In the case of a single C atom, since only photoionization and Auger decay contribute to ionization, as seen in figure 7(a), the final charge number equals to the total ionization event number. In the molecular model (figure 7(b)), a significant number of secondary ionization events occur. This increase of ionization events by secondary ionization is traded off to a certain degree by its resultant suppression of Auger decay, since the two mechanisms both involve valence electrons and therefore compete with each other (in figure 7 the overall suppression of Auger and photoionization events is seen for x-ray fluence over 1000 J cm^{-2} and can also be attributed to the deletion of core electrons by very intense x-ray pulses). As a result, with a similar number of photonization events, the total number of ionization events remains similar in both models. A dramatic deviation of the molecular model from the atomic one is that the final average atomic ion charge is 1-2 charges less than the average number of ionization per atom, indicating the occurrence of electron recombination that further suppresses ionization. For the short pulse, photoionization is suppressed similarly in both models due to core-electron depletion. However, in a molecule, with a short pulse a stronger attractive potential of the atomic ions can be built up that enhances electron trapping, and thus enhances recombination. Consequently, application of short pulse has a stronger effect in the suppression of ionization in molecules than in atoms, as indicated by the larger difference in the final charge number at short and at long pulses in the molecular model compared to the atomic model (figure 7). With all of the above, we conclude that for $C_{60,}$ the high charge suppression in the ion yields is due to electron trapping followed by recombination.

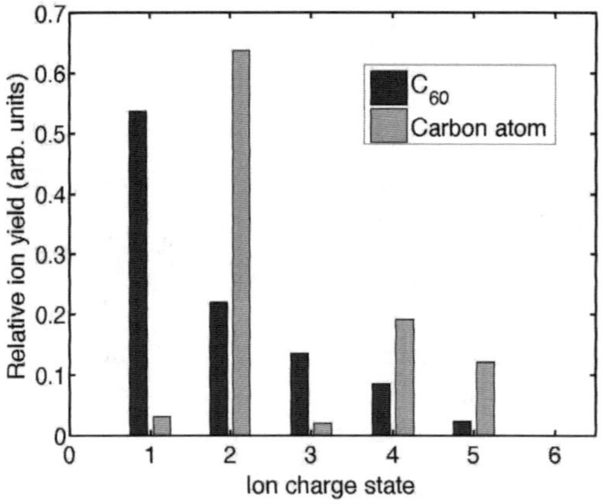

Figure 6. Theoretical ion yields for a single carbon atom and C_{60}. For this calculation the pulse duration was 30 fs, and the pulse energy was 345 μJ. The volume integrated signals were calculated using pulse parameters from the experiment. The huge discrepancy in the C^{1+} and C^{2+} yields is a direct consequence of atomic Auger (that produces C^{2+} from a neutral atom) and molecular Auger effect (that produces two C^{1+} ions from neutral ones in C_{60}), respectively. The figure is adapted from [2].

3.4. Contributions of various molecular effects in the interaction of C_{60} with intense x-rays

In addition to the various molecular dynamical processes that play distinct roles in the ionization progression, there are other molecular effects that influence the ionization, such as chemical bonds.

With our approach of adding and removing particular molecular effects in the model, we are able to determine the relative contributions of each molecular component to the final ionic outcomes. Figure 8 shows the comparison of fractional ion yields from the experiment and from the modelling with selected molecular effects included. The different degrees of discrepancy between the simulation at various conditions and the experimental data show the importance of various molecular effects in the photon interaction with C_{60}. The comparison clearly shows that i) all molecular effects are needed for the best agreement, ii) the failure of a model without any molecular effect, and iii) the importance of secondary ionization even for a midsize molecule as C_{60}. As seen in figure 8, the inclusion of the secondary ionization leads to the most significant improvement of the model to match the experimental data. Although secondary ionization has been previously reported in the case of very large Xe VdW clusters [9] and solid aluminium [10] our investigation, to the best of our knowledge, reveals for the first time significant collisional ionization *within a single molecule* exposed to intense x-rays.

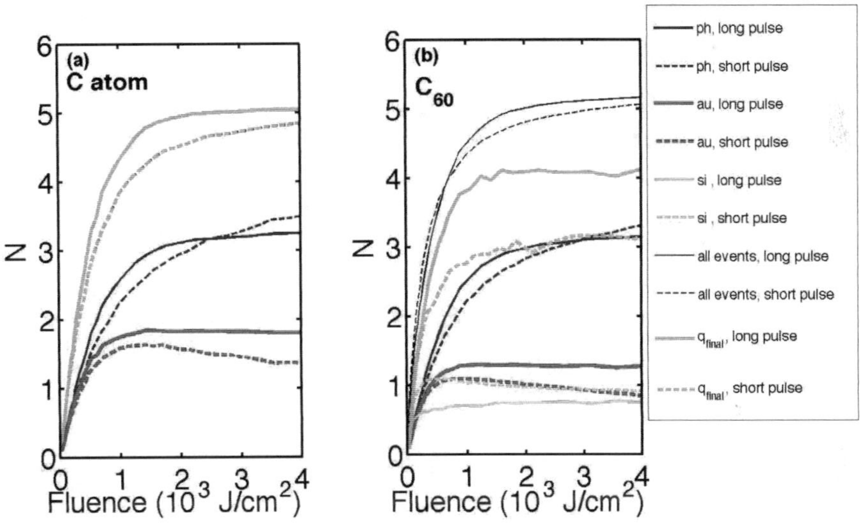

Figure 7. Comparison of atomic and molecular absorption versus volume integrated x-ray pulse peak fluence. Number of ionization events and final charge states per atom for (a) single carbon and (b) C_{60} at long and short pulse duration are shown as a function of fluence. Abbreviations: ph: photoionizations, au: Auger events, si: secondary ionization events, all: the sum of all ionization events; Q_{final}: final charge state, after recombinations in the C_{60} case. For the single carbon case there is no secondary ionization, and the "all" and "Q_{final}" curves overlap. The figure is adapted from [2].

Intuitively, one would expect that completely neglecting molecular bonds will always result in larger KEs, because we remove attractive binding energy from the system (in the C_{60} case it is approximately –430 eV). However, according to our model for the high fluence, long pulse cases, neglecting the molecular bonds can in fact have the opposite effect, as shown in figure 9(a). Without bonds the molecular ions can start expanding very early during the pulse with less chance to build up a concentrated high charge and in this way resulting in reduced ion KE. However, under certain conditions neglect of bonds can be reasonable. When the ionization is so fast the atoms have no chance to move before the bonds are broken, as in the case of short pulse and high fluence, the influence of the chemical bonds reduces. As shown in figure 9 for the 4 fs pulse duration case, bondless modelling yields practically the same KEs and ion yields in this regime (short pulse and high fluence), indicating that the simplification of neglecting chemical bonds in radiation damage calculations is adequate for fast ionization by x-ray beam of very high intensity.

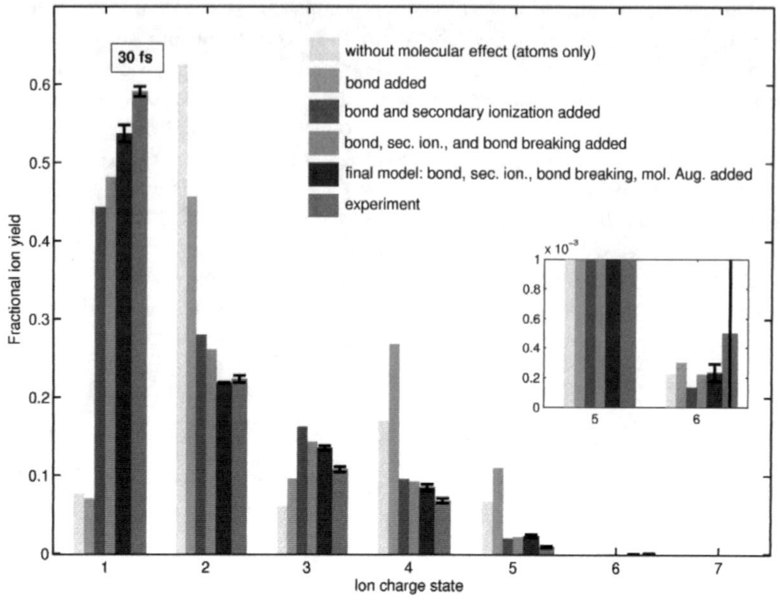

Figure 8. Importance of molecular effects reflected by the discrepancy between ion yields from experiment and from different modeling scenarios. Atomic ion yields were extracted using a model without any molecular phenomena and with an increasing number of effects until reaching the final model. Data are for the intermediate pulse duration case. Experimental data error bars are one standard deviation; modeling error bars are one standard deviation using Poisson statistics. The C^{6+} yield is near the experimental detection limit (inset). The figure is adapted from [2].

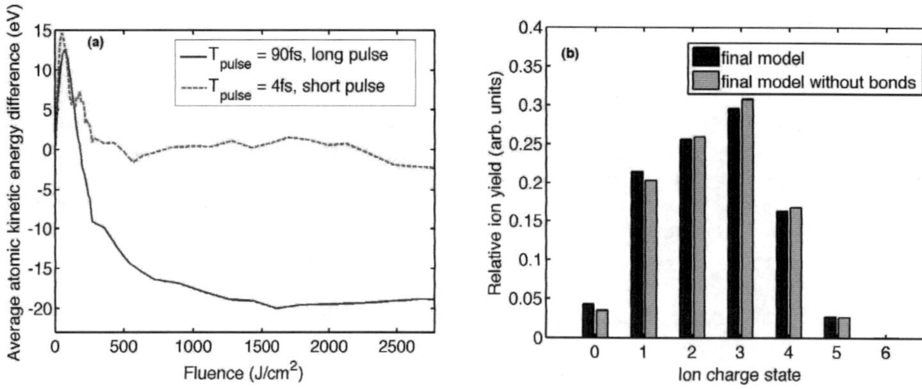

Figure 9. Modelled kinetic energy and ion charge state of atomic fragments from C_{60} as a function of fluence. The difference (a) between average atomic kinetic energies obtained from the model without bonds and that obtained from the full model ($KE^{\text{bondless model}} - KE^{\text{full model}}$) are shown for two pulse durations. Atomic ion yields (b) show agreement between the two models in the short pulse, high fluence case (in this example: photon energy is 485 eV, pulse duration 4 fs, fluence is 550 J cm^{-2}). The figure is adapted from [2].

4. Conclusion

The photo-absorption of x-rays from an FEL by a strongly bonded molecule such as C_{60} displays complex dynamics driven by various physical and chemical phenomena. With a model combining classical MD and quantum methods, we are able to investigate molecular dynamical processes and the time evolution of the molecular system. Our model was validated by good agreement between the experimental data and the simulation results. We find that various molecular mechanisms, such as secondary ionization, electron recombination and chemical bonds, play significant roles in determining the final products. The finding of these various contributions to the ionization dynamics can be used to manipulate and control a molecular system by tuning appropriately the x-ray beam parameters. Our model provides a tool to gain insight into molecular dynamics under extreme conditions resultant from the interaction with intense FEL pulses and can be extended and used for larger molecules such as bio-systems.

Acknowledgments
This work has been supported by the Department of Energy, Office of Science, BasicEnergy Sciences, Division of Chemical Sciences, Geosciences, and Biosciences, by the excellence cluster 'The Hamburg Centre forUltrafast Imaging-Structure, Dynamics and Control of Matter at the Atomic Scale' of the Deutsche Forschungsgemeinschaft.

References
[1] Berrah N *et al.* 2010 *J. Mod. Opt.* **57** 1015
[2] Murphy B F *et al.* 2014 *Nature Communication* **5** 4281
[3] Young L *et al.* 2009 *Nature* **466** 56
[4] Fang L *et al.* 2013 *Phys. Rev. Lett.* **109** 263001
[5] Bostedt C *et al.* 2013 *J. Phys. B: At. Mol. Opt. Phys.* **46** 164003
[6] Neutze R *et al.* 2000 *Nature* **406** 752; Jurek Z, Faigel G and Tegze M 2004 *Eur. Phys. J. D.* **29** 217; Hau-Riege S P, Londan R A and Szoke A 2004 *Phys. Rev. E* **69** 051906
[7] Jurek Z, Ziaja B and Santra R 2013 *XMDYN. Rev. 1.0360* (CFEL, DESY); Son S-K and Santra R 2011 *XATOM*—an integrated toolkit for x-ray and atomic physics (CFEL, DESY)
[8] Berrah N *et al. Faraday Discussions 2014* DOI: 10.1039/C4FD00015C
[9] Thomas H *et al.* 2012 *Phys. Rev. Lett.* **108** 133401
[10] Ciricosta O *et al.* 2012 *Phys. Rev. Lett.* **109** 065002

MPS2014 IOP Publishing
Journal of Physics: Conference Series **601** (2015) 012007 doi:10.1088/1742-6596/601/1/012007

Single and Double Ionization of Atoms and Small Molecules by Short-Pulse Intense Laser Fields

Klaus Bartschat
Department of Physics and Astronomy, Drake University, Des Moines 50322, USA
E-mail: klaus.bartschat@drake.edu

Abstract. Examples of recent calculations for the interaction of intense short-pulse laser fields with (quasi)one- and two-electron atomic (H, He, Li) and molecular (H_2^+, H_2) targets are presented. In light of the often challenging parameter sets chosen in both experiment and other theoretical work, the need to carefully check both the numerical aspects and physical effects such as resonances is emphasized.

1. Introduction

The response of atoms and molecules irradiated by intense laser pulses has attracted growing interest in recent years (see, for example, [1] for a recent collection of papers). This is to a large extent due to the rapid progress in both experimental and theoretical/computational techniques regarding few-cycle infrared (IR) pulses, free-electron lasers (FELs) and high-harmonic generation (HHG) that can provide extreme ultraviolet (XUV) pulses, and super-computer resources that allow for the direct solution of the time-dependent Schrödinger equation (TDSE) on a space-time grid using massively parallel machines. For (quasi)-one electron targets, very detailed information about the response of the active electron to single and multiple (e.g. in pump-probe setups) laser pulses can be obtained, while the correlated response of multi-electron targets to strong driving laser fields can also be studied in both the time and momentum domains.

Over the past few years, our group at Drake University has been engaged in performing a number of benchmark calculations for simple atoms and molecules, such as H, He, Li, H_2^+ and H_2. An overview of our work (not a review of the field) with a few selected examples is presented in this paper. Work on heavier targets (Ne and Ar) can be found in refs. [2, 3]. Despite the apparent simplicity of these targets, it turns out that some of the parameters (central frequency, peak intensity, pulse length, pulse shape, etc.) chosen in many experiments, as well as other theoretical studies, make the calculations extremely challenging. Hence, we have put strong emphasis on developing accurate numerical methods and the accompanying computer programs. In rather extreme scenarios, somewhat surprising results were obtained. However, the extensive tests carried out give us confidence in our predictions.

In the following section, we briefly describe some of the numerical techniques implemented in our programs. This is followed by a few illustrative examples. Details can be found in the references given. Unless indicated otherwise, atomic units are used throughout this paper.

2. Numerical Methods

The results presented below were obtained with a number of different approaches. For H and Li, we employed the matrix iteration method (MIM) of Nurhuda and Faisal [4] to propagate the initial field-free state in time under the influence of a strong laser field. Depending on the peak intensity of the pulse(s), we use either the length or the velocity form of the electric dipole operator. Details were described by Grum-Grzhimailo *et al* [5]. The method is very efficient and numerically accurate. However, it requires the solution of a tridiagonal system of equations at every timestep, both in the radial coordinate (when using a finite-difference method) and in the angular momentum due to the dipole coupling of different partial-wave symmetries. Hence, special care has to be taken to ensure efficient parallelization of the computer code.

For He, we combined field-free and dipole matrices obtained with the B-spline R-matrix method [6, 7] with an efficient short-iterative Lanczos (SIL) [8] propagation scheme. This is an explicit method that effectively reduces to matrix-vector multiplications, for which highly optimized library routines are widely available. The computer code used for single ionization of Ne and Ar is published [9]. Such programs run efficiently on passively parallel machines using a hybrid technique of OpenMP and MPI parallelization.

Finally, for H_2^+ and H_2, we again employed the SIL technique, but this time combined with a finite-element discrete-variable representation (FE-DVR). Due to the lack of spherical symmetry and the entangled motions of the nuclei and the electrons, the situation in molecules is much more complicated than in atoms. The latter complication is often neglected by limiting the treatment to the fixed-nuclei approximation (FNA), which in turn limits the comparison with experiment. Even though the electron-electron interaction (needed for the H_2 problem) is complicated, formulating the problem in prolate-spheroidal coordinates and thereby exploring the natural symmetry of the problem is highly advantageous.

3. Example Results

This section provides an overview of some of the calculations that have been performed at Drake University over the past few years. As mentioned above, strong emphasis has been put on the numerical stability of the results, including those from essentially model problems for which experimental do not (yet) exist. We start with atomic hydrogen and lithium, assuming that the latter can effectively be described as a quasi-one-electron target with the inner two electrons being represented by a core potential. We then move on to helium as a two-electron target before going to the H_2^+ and H_2 molecules.

3.1. H and Li

Figure 1 shows an example for ionization of atomic hydrogen by a few-cycle 780 nm laser pulse of peak intensity of 10^{15} W/cm^2. Handling such a strong IR field far outside the perturbative regime represents a challenging numerical problem. Using the velocity gauge of the electric dipole operator, we could demonstrate the stability of our predictions against variations of various parameters, such as the number of coupled partial waves. Details can be found in [5].

The above work laid the foundations for several follow-up projects, including the necessary assistance from theory in calibrating absolute laser intensities at the 1% level [10]. It also led to the prediction of a displacement effect [11], i.e., a strong dependence of the photoelectron spectrum and the angular distribution on what on first sight seemed to be very small changes in the ramp-on and ramp-off characteristics of a laser pulse with a plateau of the envelope function in between. Figure 2 shows an example of two pulses denoted by "2-36-2 S-S" and "2-36-2 L-L", respectively. Here "n_1-n_2-n_3" refers to the number of optical cycles (o.c.) in the ramp-on (n_1), the plateau (n_2), and the ramp-off (n_3), while "S" and "L" label sine-squared (S) or linear (L) ramp-on/off.

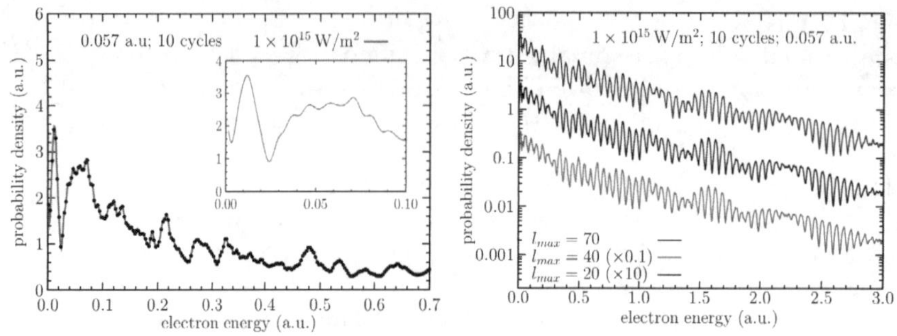

Figure 1. Photoelectron spectrum for ionization of atomic hydrogen in a 10-cycle laser pulse with a $\sin^2$ envelope for the vector potential, a peak intensity of $10^{15}\,\mathrm{W/cm^2}$, and a central frequency of 0.057 a.u., corresponding to a wavelength of 780 nm. [5]. Left: Results obtained with the velocity form of the electric dipole operator for angular momenta up to $\ell_{max} = 70$ (solid line) and $\ell_{max} = 50$ (dots). Right: Similar results over an extended energy scale. The results for $\ell_{max} = 40\,(20)$ were multiplied by $0.1\,(10)$ to make them distinguishable from those for $\ell_{max} = 70$.

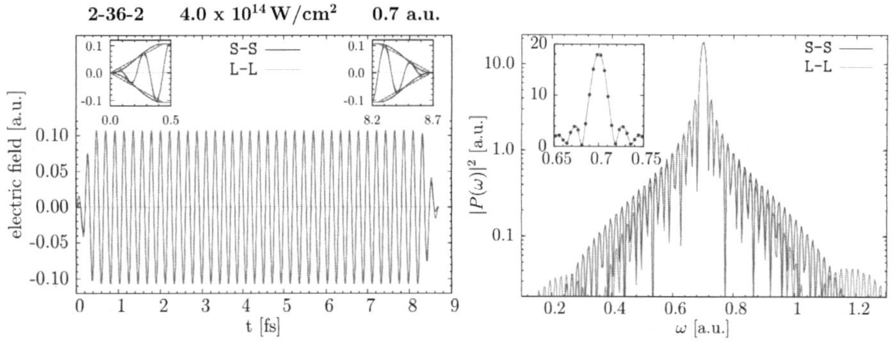

Figure 2. Electric field (left) and Fourier spectrum (right) of 2-36-2 L-L and S-S pulses with central photon energy 0.7 a.u. and peak intensity 4.0×10^{14} W/cm^2 [11]. The inserts magnify the changes due to the small differences in the ramp-on/off cycles. The pulses are identical in the plateau regime.

Because of its universal nature, the displacement effect should be observable in any atom or molecule – provided pulses with non-zero displacement can be made experimentally. This question is currently under discussion among experimental experts, but the existence of such pulses *per se* does not seem to contradict Maxwell's equations. Indeed, Fig. 3 displays ionization spectra and photoelectron angular distributions (PADs) for Li driven by a similar set of S-S and L-L pulses. The ramp-on/off effect in the energy spectra is very similar to that observed for hydrogen [11]. It manifests itself in the PADs integrated over the energy interval covering approximately half of the ionization peak in Fig. 3, while it essentially disappears if a symmetric energy window is used.

The left part of Fig. 4 illustrates the momentum distribution of the electron ejected from Li by a few-cycle single laser pulse of peak intensity $\approx 4 \times 10^{11}$ W/cm^2 and a sine-squared envelope of the electric field [12]. Relating it to the right part, where two pulses of peak intensity $\approx 10^{11}$ W/cm^2 delayed by 170 femtoseconds (fs) act on the system, a single pulse corresponds

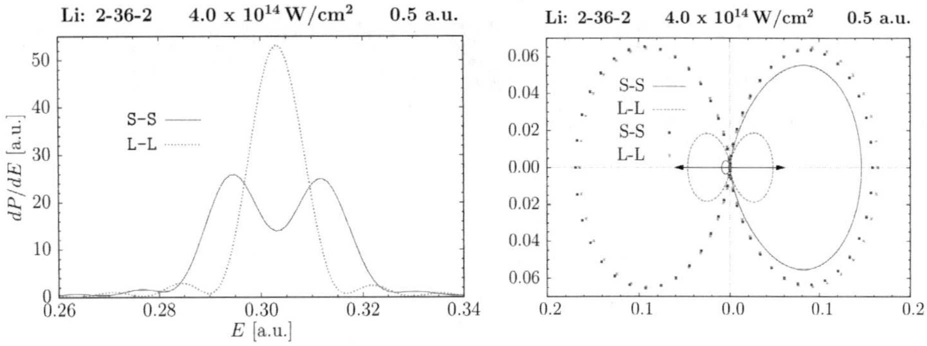

Figure 3. Left: Ejected-electron spectrum for ionization of Li by 2-36-2 S-S (solid line) and 2-36-2 L-L (dashed line) pulses with central photon energy 0.5 a.u. and peak intensity 4.0×10^{14} W/cm^2 [11]. Right: PADs for Li by pulses with central photon energy 0.5 a.u. and peak intensity 4.0×10^{14} W/cm^2. Lines are for an asymmetric energy window, 0.25 a.u. $\leq E \leq$ 0.30 a.u., while symbols are for a symmetric energy window, 0.25 a.u. $\leq E \leq$ 0.35 a.u., around the central peak. The arrow indicates the direction of the laser polarization axis.

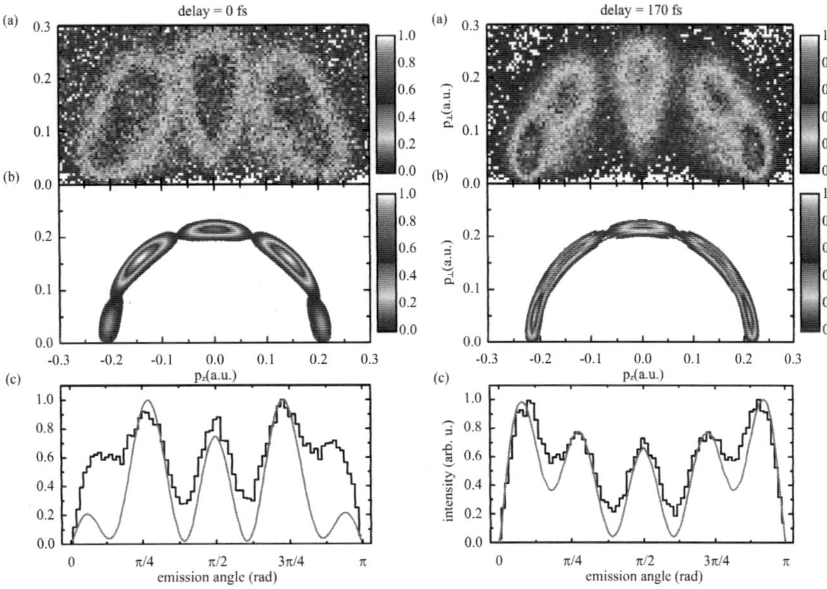

Figure 4. Left: Momentum distribution of the ejected electron after a single linearly polarized laser pulse of peak intensity $\approx 4 \times 10^{11}$ W/cm^2 as observed in experiment (a) and theory (b) [12]. The respective peak normalized angular distributions (black: experiment; red: theory) are shown in (c). Right: Momentum distribution of the ejected electron for a delay of 170 fs between two linearly polarized laser pulses of 30 fs (FWHM) duration, as observed in experiment (top row) and theory (center row). The peak-normalized PADs are shown in the bottom row.

to two such pulses without delay. In the experimental momentum distribution, five angular maxima are clearly identified. The experimental momentum resolution of about 0.05 a.u. in longitudinal direction and 0.07 a.u. in transversal direction gives rise to a broadening of the

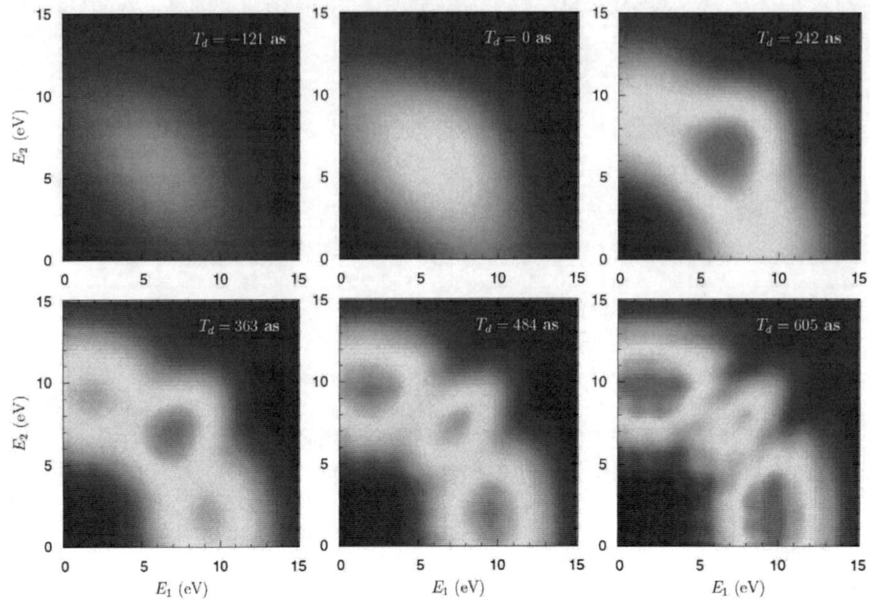

Figure 5. Energy distributions of the two escaping electrons from He in a two-color laser pulse scenario [14]. The laser parameters are: $\omega_1 = 35.3$ eV, $\tau_1 = 12$ o.c. at a peak intensity of $10^{14}\,\mathrm{W/cm^2}$ and $\omega_2 = 57.1$ eV, $\tau_2 = 14$ o.c. at a peak intensity of $10^{13}\,\mathrm{W/cm^2}$. The time delays between the two laser pulses are indicated in the panels.

peaks shown in the top row of Fig. 4 compared to the theoretical results in the middle row. A more quantitative comparison between experiment and theory is shown in the bottom row, where the PAD is displayed for the main photoline.

3.2. He

Double-ionization (DI) of He as the classic two-electron target is another highly challenging problem, due to the correlated behavior of the two electrons both in the initial and the final state consisting of three charged particles moving away from each other. One of the many issues concerns the extraction of the information after the laser field is switched off [13]. In one of our works on this particular problem [14], we investigated a two-color pump-probe scheme, in which two laser pulses with a tunable delay hit the atom. The central frequencies are chosen in such a way that one of the pulses could directly only eject an electron from the He($1s^2$) ground state (ionization potential of 24.6 eV), while the other one could also eject the remaining electron from the He(^+1s) ground state, which requires 54.4 eV of energy. Hence one would expect a non-sequential process if the pulse with the higher frequency comes first, while a sequential process appears possible otherwise. However, the latter is not obvious, since the outcome may also depend on the relaxation time of the singly-ionized state, i.e., the time it takes for the remaining electron after the first ionization to realize that it is in the "wrong" quantum state.

Figure 5 shows the energy distribution of the two escaping electrons for various delays between two pulses with central frequencies corresponding to 35.3 and 57.1 eV, respectively. A negative delay means that the higher-frequency pulse comes first. As seen from the figure, it takes several hundred attoseconds (as) before the sequential process is clearly established, i.e., the expected energies of two subsequently electrons are predicted with high probability.

Journal of Physics: Conference Series **601** (2015) 012007 doi:10.1088/1742-6596/601/1/012007

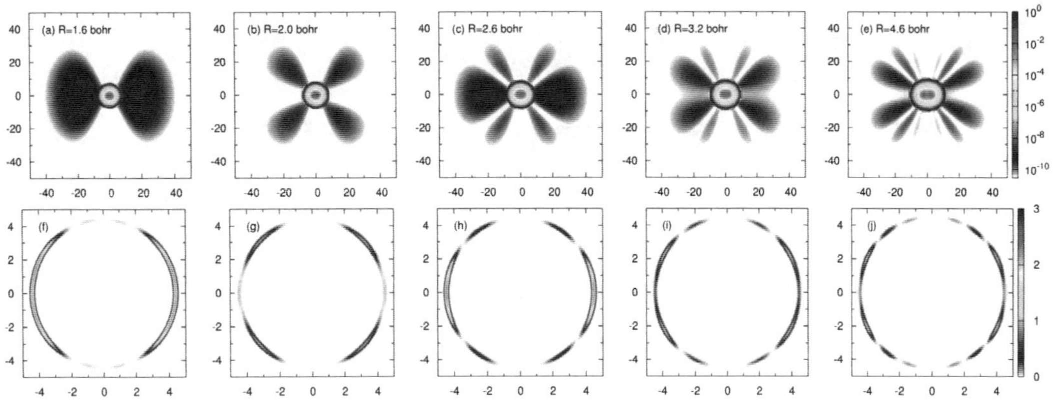

Figure 6. Probability densities (top) and momentum distributions (bottom) of the electron in the H_2^+ ion due to a 20-cycle $\sin^2$ pulse with central energy of 300 eV and peak intensity 10^{16} W/cm^2 in the parallel (horizontal in these panels) geometry [15].

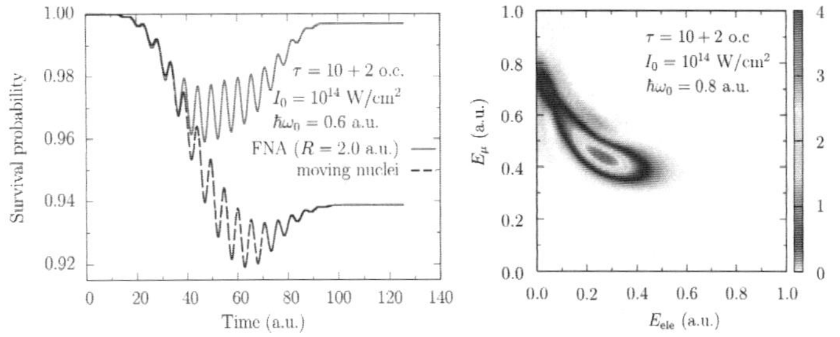

Figure 7. Left: Survival probability as a function of time for the ground state of the H_2^+ molecule exposed to a 10-cycle laser pulse ($\sin^2$ envelope) of central energy 0.6 a.u. and intensity 10^{14} W/cm^2 in the parallel geometry. "10+2 o.c." indicates that the information is extracted after another two optical cycles of field-free propagation of the wave packet. Right: Sharing of the excess energy between the nuclei and the electrons [18].

3.3. H_2^+

Moving on to molecules, two immediate complications arise. First, the multi-center nature of the target allows for the study of orientation effects due to the relative directions of the linear laser polarization vector and the internuclear axis. Second, the nuclei will move in reality, but even in the FNA one can study potential interference effects, i.e., the results obtained with different values of the internuclear distance R may vary substantially, since it is quantum mechanically impossible to assign which nucleus the electron came from.

Figure 6 displays the probability densities in configuration space and the momentum distributions at the end of the laser pulse, respectively, in the parallel geometry, where the electric field of the laser and internuclear axis are oriented along the same direction. Though the time-evolution of the wave packet in configuration space does not directly reveal the dominant momentum distribution of the photoelectron, it clearly shows both the preferred and forbidden emission directions. For some cases, the probability density for the electron ionized along the polarization axis is completely negligible [c.f. Fig. 6 (c) at $R = 2.0$]. This corresponds to the "confinement effect" [16] observed in the angular-resolved momentum distribution [c.f. Fig. 6 (g)]. Similar correspondences in the perpendicular geometry can also be established [15].

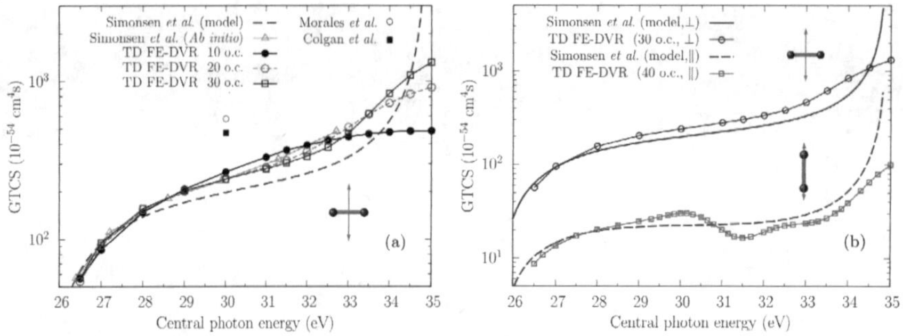

Figure 8. GTCS for two-photon DI of the H_2 molecule. In panel (a), the molecular axis is oriented perpendicular to the linear polarization vector (represented by a double-headed arrow) of the laser pulse. Our results were obtained for pulse lengths of 10, 20, and 30 optical cycles. Also shown are the *ab initio* and the model results of Simonsen *et al.* [21], as well as the cross sections of Colgan *et al.* [19] and Morales *et al.* [20] at 30 eV. Panel (b) shows a comparison between our *ab initio* results obtained in the perpendicular ($\perp$) and parallel ($\|$) geometries, and again the corresponding predictions from the simple model.

As mentioned previously, limiting the theoretical model to the FNA usually prevents direct comparison with experiment. For H_2^+, indeed, this restriction can be relaxed [17]. Figure 7 shows preliminary results [18] for the survival probability of the H_2^+ ground state (left panel) and the predicted sharing of the excess energy between the nuclei (E_μ) and the ejected electron ($E_{\rm ele}$) (right panel) when the H_2^+ molecular ion is exposed to a pulse of central energy 0.6 a.u. or 0.8 a.u., respectively. The dominant process here is two-photon ionization. The FNA results clearly change when the nuclei are allowed to move. Hence, it is essential to treat this process in a more general way than in the FNA.

3.4. H_2

Finally, molecular hydrogen as a two-center two-electron problem remains a formidable challenge to theory. Several *ab initio* calculations for this problem have been performed, using highly sophisticated methods such as time-dependent close-coupling (TDCC) [19] and exterior complex scaling (ECS) [20]. Unfortunately, the results for apparently similar parameters do not always agree with each other to the extent that one might have expected.

The angle-integrated generalized total cross sections (GTCSs) for double ionization of the H_2 molecule for various cases are displayed in Fig. 8. At $h\nu = 30$ eV, all previously published time-dependent calculations [19, 21, 22] employed an interaction time between 1.4-2.1 fs, i.e., 10-15 o.c. For the perpendicular geometry shown in Fig. 8 (a), the dependence of our results on the pulse length is small for central photon energies up to about 33 eV, as shown by examining predictions obtained with 10, 20, and 30 o.c. When approaching the threshold for sequential DI, on the other hand, the rapid increase of the GTCS is not at all captured by the 10 o.c. pulse. Our results for 20 o.c. are in excellent agreement with the *ab initio* calculations by Simonsen *et al.* [21], which were performed using an entirely independent method, but with similar laser parameters (15 o.c. as the pulse duration). The simplified model of Ref. [21] also does reasonably well. The time-dependent close-coupling (TDCC) results of Colgan *et al.* [19] and the exterior complex scaling (ECS) predictions of Morales *et al.* [20] for a photon energy of 30 eV, however, are about a factor of two larger than our TDSE results at this energy. In contrast to the parallel geometry where even larger discrepancies occur, resonance effects due to autoionizing states, which may affect the FNA results [22, 23], can be ruled out for the perpendicular case.

4. Conclusions and Outlook

We have presented a number of benchmark calculations for intense short-pulse laser interactions with small atoms and molecules. For the atoms, the TDSE can be solved with high accuracy and hence it is possible to reproduce experimental data in detail and also guide experiment where appropriate. Some of the experimental parameters chosen are computationally challenging, and hence great care must be taken to ensure numerical accuracy. For molecules, most calculations remain limited to the fixed-nuclei approximation, which makes a meaningful comparison with experiment difficult, if not impossible. As an important intermediate step towards the ultimate goal of accounting for the nuclear motion, calculations for a number of well-defined scenarios should be carried out by independent groups, and the results should be compared carefully.

Acknowledgments

This work was supported by the United States National Science Foundation under grants in the Atomic Theory and Computational Physics programs, and through the eXtreme Science and Engineering Discovery Environment. I am most grateful to Drs. X. Guan, A. N. Grum-Grzhimailo, and O. Zatsarinny, who developed the computer programs used for these calculations. I am also indebted to many theoretical collaborators, including B. Abeln, S. Buczek, R. DuToit, J. Emmons, J. Feist, E. Gryzlova, I. Ivanov, A. Kheifets, L. Koesterke, C. Noble, B. Schneider, E. Secor, T. Urness, and D. Weflen, as well as the experimental groups of A. Dorn, R. Mooshammer, and J. Ullrich at the Max-Planck Institut for Nuclear Physics in Heidelberg and of D. Kielpinski, I. Litvinyuk, and R. Sang at Griffith University in Brisbane.

References

[1] Martín F, Hishikawa A and Vrakking M 2014, *Special issue on ultrafast electron and molecular dynamics J. Phys. B* **47**, entire issue #12 (June)
[2] Guan X, Zatsarinny O, Bartschat K, Schneider B I, Feist J and Noble C J 2007 *Phys. Rev. A* **76** 053411
[3] Guan X, Noble C J, Zatsarinny O, Bartschat K and Schneider B I 2008 *Phys. Rev. A* **78** 053402
[4] Nurhuda M and Faisal F H M 1999 *Phys. Rev. A* **60** 3125
[5] Grum-Grzhimailo A N, Abeln B, Bartschat K, Weflen D and Urness T 2010 *Phys. Rev. A* **81** 043408
[6] Zatsarinny O 2006 *Comp. Phys. Commun.* **174** 273
[7] Zatsarinny O and Bartschat K 2013 *J. Phys. B* **46** 112001
[8] Park T J and Light J C 1986 *J. Chem. Phys.* **85** 5870
[9] Guan X, Noble C J, Zatsarinny O, Bartschat K and Schneider B I 2009 *Comp. Phys. Commun.* **180** 2401
[10] Pullen M G, Wallace W C, Laban D E, Palmer A J, Hanne G F, Grum-Grzhimailo A N, Bartschat K, Ivanov I A, Kheifets A S, Wells D, Quiney H M, Tong X M, Litvinyuk I V, Sang R T, and Kielpinski D 2013, *Phys. Rev. A* **87** 053411
[11] Ivanov I A, Kheifets A S, Bartschat K, Emmons J, Buczek S M, Gryzlova E V and Grum-Grzhimailo A N 2014 *Phys. Rev. A* **90** 043401
[12] Schuricke M, Bartschat K, Grum-Grzhimailo A N, Zhu G, Steinmann J, Moshammer R, Ullrich J and Dorn A 2013 *Phys. Rev. A* **88** 023427
[13] Guan X, Bartschat K, and Schneider B I 2008 *Phys. Rev. A* **77** 043421
[14] Guan X, Zatsarinny O, Noble C J, Bartschat K and Schneider B I 2009 *J. Phys. B* **42** 134015
[15] Guan X, Secor E B, Bartschat K and Schneider B I 2012 *Phys. Rev. A* **85** 043419
[16] Fernández J, Fojón O, Paciaos A and Martín F 2007 *Phys. Rev. Lett.* **98** 043005
[17] Hu S X, Collins L A and Schneider B I 2009 *Phys. Rev. A* **80** 023426
[18] Guan X 2014 *private communication*
[19] Colgan J, Pindzola M S and Robicheaux F 2008 *J. Phys. B* **41** 121002
[20] Morales F, Martín F, Horner D A, Rescigno T N and McCurdy C W 2009 *J. Phys. B* **42** 134013. The unpublished total cross section was obtained via private communication from Dr. Morales.
[21] Simonsen A S, Sørngård S A, Nepstad R and Førre M 2012 *Phys. Rev. A* **85** 063404
[22] Guan X, Bartschat K, and Schneider B I 2010 *Phys. Rev. A* **82** 041404(R)
[23] Guan X, Bartschat K, Schneider B I, and Koesterke L 2013 *Phys. Rev. A* **88** 043402

Charge transfer processes in proton-helium collisions: The validity of the first Born approximation

Yu. V. Popov[1,3], A. Galstyan[1,2], O. Chuluunbaatar[3,4], S. Houamer[5], A. A. Bulychev[3], M. S. Schöffler[6], H.-K. Kim[6], J. N. Titze[6], T. Jahnke[6], L. Ph. H. Schmidt[6], H. Schmidt-Böcking[6], R. Dörner[6]

[1]Skobeltsyn Institute of Nuclear Physics, Lomonosov Moscow State University, Moscow 119991, Russia
[2]Faculty of Physics, Lomonosov Moscow State University, Moscow 119991, Russia
[3]Joint Institute for Nuclear Research, Dubna, Moscow region 141980, Russia
[4]School of Mathematics and Computer Science, National University of Mongolia, UlaanBaatar, Mongolia
[5]Département de physique, Faculté des Sciences, Université Ferhat Abbas, Sétif, 19000, Algeria
[6]Institut für Kernphysik, University Frankfurt, Max-von-Laue-Str. 1, 60438 Frankfurt, Germany

E-mail: popov@srd.sinp.msu.ru

Abstract. The validity of the Born series expansion for the charge transfer reactions is studied in the case of a proton-helium collision. Three different channels are considered, namely the charge transfer, transfer excitation and transfer ionization. The differential cross sections and the contributions from different charge transfer mechanisms within various Born approximations are compared with experimental data. The role of the electron-electron correlations in the initial helium state is discussed in detail. It is shown that the first Born approximation is valid in the case of reactions under consideration, provided very small scattering angles are involved and the proton energy is >500 keV. It is also shown that the electron-electron correlations in the initial helium state are important only in transfer excitation and transfer ionization reactions.

1. Introduction

The Born series expansion is a powerful tool for studying a wide range of fast scattering processes in atomic physics. There is a common belief that the first Born approximation is not valid in the case of charge transfer processes [1]. This is because the major contribution comes from the second- and higher-order terms of the series. However, it seems strange that the approximation perfectly applicable for other scattering problems completely fails when describing charge transfer.

Charge transfer reactions, at impact energies ranging from eVs in microbiology to MeVs, or even GeVs, in astrophysics, are a very common channel of energy deposition. One of the most simple charge transfer reactions is the proton impinging on a helium atom. This problem is not fully studied even after hundred years since the quantum mechanics formulation emerged. It can end up with three different channels:

- Charge transfer (CT)
  $p + \mathrm{He} \to \mathrm{H} + \mathrm{He}^+$

- Transfer excitation (TE)
  $p + \mathrm{He} \rightarrow \mathrm{H} + \mathrm{He}^{+*}$
- Transfer ionization (TI)
  $p + \mathrm{He} \rightarrow \mathrm{H} + e^- + \mathrm{He}^{++}$

All these channels are considered in the present contribution.

In the plane wave first Born approximation (PWFBA) these reactions can be considered from the viewpoint of two main mechanisms, namely the shake-off (SO) and binary encounter (BE) mechanisms [2]. Contributions of these different mechanisms can be treated separately even within the PWFBA.

Another, still open question is the role of electron-electron correlations. Many publications are devoted to this problem (see, for example, the classical textbook [3]), but some powerful details of the influence of electron-electron correlations in the initial state on different channels of charge transfer reactions remain still not well understood.

In this short contribution we show that the first Born approximation appears to be still applicable in the case of CT processes. However, its use implies limitations on the kinematical conditions, namely scattering angles and impact energies. Furthermore, we show that the influence of the electron-electron correlation in the initial helium state plays an important role in the cases of TI and TE reactions, and seems to be subsidiary in the CT case.

2. Theory

The Hamiltonian for the CT reaction can be written either in the *prior*-, either in the *post*-form. In the *prior*-form it looks like

$$H_i = [H_{He} + h_p] + V_{p1} + V_{p2} + V_{pN}, \tag{1}$$

and its *post*-form is

$$H_f = [H_H + H_{He^+}] + V_{p2} + V_{pN} + V_{N1} + V_{12} \tag{2}$$

Here "1" and "2" label electrons, "p" the incident proton, "H" the final moving atomic hydrogen in its ground state (the experiment [4] allows to distinguish its ground and excited states), "He" the helium atom in its ground state, "He$^+$" the final helium ion in an arbitrary quantum state including the continuum state $(N + e)$, "N" the helium nucleus. The operators in the square brackets define the asymptotics of the in (out) channel of this reaction, the potential terms $\sum V_i$ are the corresponding perturbations. More details and the integral representation of the corresponding Schrödinger equations for (1) and (2) can be found in [4].

The matrix element in the *prior*-form follows from (1):

$$T_{fi} = \langle \Psi_f^- | V_{p1} + V_{p2} + V_{pN} | \Psi_i^{as} \rangle. \tag{3}$$

The PWFBA (also known as the Jackson-Schiff approximation [5]) can be deduced from (3) if one uses the initial and final states in their asymptotic forms, that is,

$$|\Psi_i^{as}\rangle = |\Phi_{He} > |\vec{p}_0\rangle$$

and

$$\langle \Psi_f^- | = \langle \varphi_H, \vec{p}_f | \langle \phi_{He^+} |.$$

Here $|\vec{p}_0\rangle$ describes the plane wave of the incident proton, $\langle \varphi_H, \vec{p}_f|$ corresponds to the hydrogen atom, whose center-of-mass moves with the momentum $\vec{p}_f$, and $\langle \phi_{He^+}|$ describes the wave function of the final helium ion, which can be in any bound or even continuum state with the momentum of an ejected electron $\vec{k}$. The PWFBA diagrams of the scattering process for the last case (transfer ionization) are presented in Fig. 1.

Two important remarks.

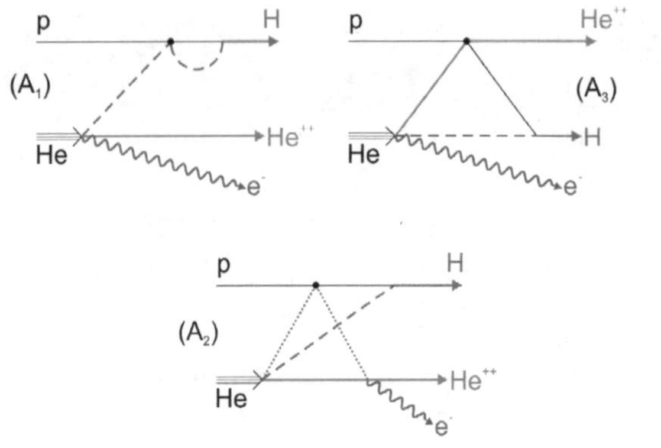

Figure 1. PWFBA diagrams for transfer ionization. The diagram A_1 (V_{p1} in (3)) is the well known Oppenheimer-Brinkman-Kramers approximation. The diagram A_3 (V_{pN} in (3)) describes the interaction of heavy particles. Both A_1 and A_3 correspond to the shake-off mechanism of emission. The diagram A_2 (V_{p2} in (3)) describes the successive emission mechanism (BE).

(i) We consider a fast projectile proton with the energy higher than 0.5 MeV and very small scattering angles θ_p, less than 1 mrad (the momentum transfer $\vec{q} = \vec{p}_f - \vec{p}_0$ is about a few atomic units). These kinematical conditions allow to consider the nucleus as being at rest because of huge masses of the proton and nucleus. In other words, we neglect in calculations such ratios as $K_{ion}^2/M_N, q^2/2m_p$.

(ii) The first Born approximation (FBA) form (3) is valid only if the initial and final asymptotic channels of the transfer reactions are not charged (see details in [6, 7]). This is realized in the present case because the He atom and the final hydrogen atoms are neutral.

For the initial state of the helium atom we use the following trial wave functions:

- The strongly correlated wave function (named as Ch) from [8]. $E_0^{Ch} = -2.9037$ a.u.
- The configuration interaction (CI) wave function of Mitroy et al. [9] with $E_0^{CI} = -2.9031$ a.u.
- A simple correlated wave function SPM from [10] with $E_0^{SPM} = -2.8952$ a.u.
- $1s^2$ Roothan-Hartree-Fock [11] wave function RHF describing two independent electrons in an averaged core electromagnetic field. $E_0^{RHF} = -2.8617$ a.u..

The experimental value of the He ground-state energy is $E_0^{exp} = -2.9037...$ a.u. Let us remind that the outgoing hydrogen atom is in its ground state.

We also calculate for comparison the eikonal wave Born approximation (EWBA), which includes the phase-factor in the matrix element (3) (see details in [4]), and the plane wave second Born approximation (PWSBA) in the CT case

$$T_{fi}^{PWSBA} = T_{fi}^{PWFBA} + \langle \varphi_H, \vec{p}_f; \phi_{He^+}|(V_{p1} + V_{p2} + V_{pN}) \times$$

$$(E - H_H + H_{He^+} + i0)^{-1}(V_{p2} + V_{pN} + V_{N1} + V_{12})|\Phi_{He}; \vec{p}_0\rangle. \quad (4)$$

In the PWSBA calculations we involved the so-called closure approximation (CA), which is usually valid for high impact energies.

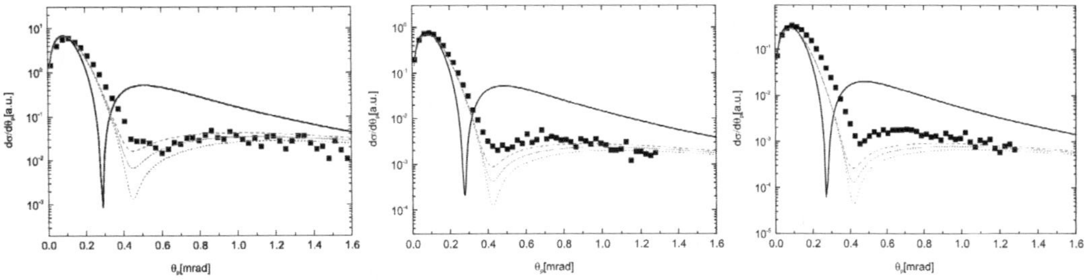

Figure 2. SDCS vs the scattering angle θ_p. Solid line is PWFBA with the RHF wave function, dashed line is PWSBA with the RHF (long dash, $\tilde{E} = 0.1$; short dash, $\tilde{E} = 0.5$; dots, $\tilde{E} = 1$). The hydrogen atom and helium ion are in their ground states. On the lhs, the incident proton energy is 630 keV, in the centre 1000 keV, on the rhs 1200 keV.

3. Results and discussion

3.1. Charge transfer

In the CT case we consider the single differential cross section (SDCS)

$$\frac{d\sigma}{d\theta_p} = 2\frac{m_p^2\theta_p}{(2\pi)}|T_{fi}|^2. \tag{5}$$

One can see in Fig. 2 that the experimental angular distribution can be split into two parts: (i) the peak structure around $\theta_p \approx 0.1$ mrad and (ii) the plateau at $\theta_p \geq 0.4$ mrad. The PWFBA calculation behaves similarly, but the main peak is more sharp and the magnitude of the cross section at larger angles, after the minimum, is ten times larger than the experimental one. Also it is worth mentioning that the shape and the magnitude of the PWFBA cross section practically does not depend on the electron correlations in He. The single differential cross section (SDCS) is mainly determined by the $1s^2$ part of the wave function.

From Fig. 2 it can be seen that the PWFBA well describes experiment in a narrow vicinity of the main peak. The PWSBA amplitude contains summation (integration) over all intermediate states of the hydrogen atom and He^+ ion. To simplify this summation we employed the CA. Here $\tilde{E}$ is the closure parameter. The PWSBA-CA calculations using the RHF wave function are shown in Fig. 2 with dashed lines. PWSBA calculations approach the PWFBA ones near the peak maximum, considerably exceeding them at larger angles, and describe the plateau much better. We must note that the experiment [4] is normalized to the PWFBA peak, which in turn gives a reasonable total cross section [12].

3.2. Transfer excitation

The single differential cross section for TE reactions is the same, but involves summation over possible final excited shells:

$$\frac{d\sigma}{d\theta_p} = 2\frac{m_p^2\theta_p}{(2\pi)}\sum_{n=2}^{n-1}\sum_{l=0}^{l}\sum_{m=-l}|T_{nlm}|^2. \tag{6}$$

In this infinite sum we keep only two states, namely with $n = 2, 3$. As shown in [13], it is sufficient to describe $\approx 90\%$ of all He^+ excited states contributions.

In Figs. 3 and 4, the experimental data and theoretical PWFBA calculations for TE are presented [13]. The absolute SDCS for TE reactions is about 5% of those for CT reactions. The PWFBA profile for the CH wave function that includes electron-electron correlations, is similar

Journal of Physics: Conference Series **601** (2015) 012008 doi:10.1088/1742-6596/601/1/012008

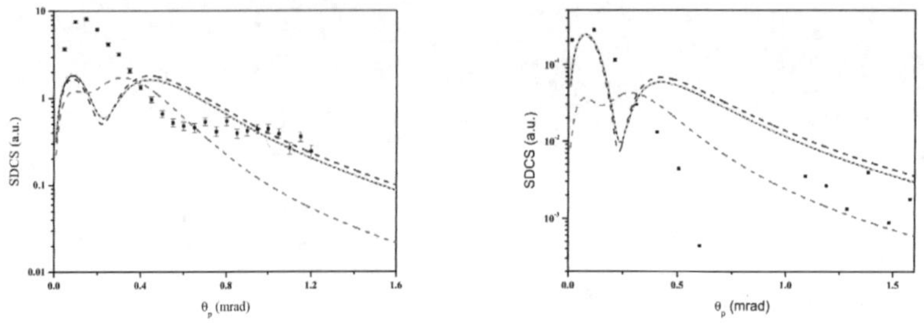

Figure 3. Experimental [13] and theoretical PWFBA data for the TE reaction. $E_p = 300$ keV on the lhs, $E_p = 630$ keV on the rhs. The hydrogen atom is in its ground state. Black squares represent experimental data, dashed red line the SDCS using the RHF helium wave function, dash-dotted blue line the SPM wave function, dotted black and solid green lines correspond to the strongly correlated helium wave functions (Ch and CI correspondingly) and practically coincide.

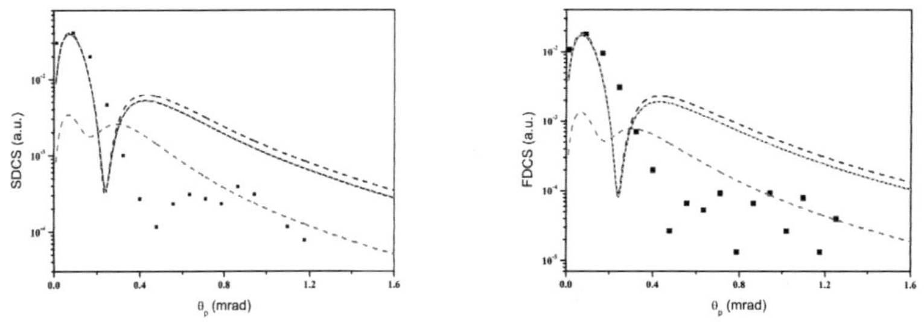

Figure 4. The same as in Fig. 3, but $E_p = 1000$ keV on the lhs and $E_p = 1200$ keV on the rhs.

to that for the CT reaction, but has the finite non-zero minimum. The calculation using the RHF wave function without angular correlations completely fails to explain the experimental data. However, as was shown earlier [4], any helium wave function describes quite well the main scattering peak for CT reactions (even the simple $1s^2$ wave function).

One can see from Fig. 4 that $E_p \sim 500$ keV is the boundary when PWFBA still satisfactorily explains the main peak at small θ_p. Also, for all impact energies, PWFBA does not describe the minimum and the plateau following this minimum at larger angles. To explain this we need SBA or EWBA calculations.

In Fig. 5, the EWBA calculations using the 9D code [13] and the SPM wave function are presented for the energies $E_p = 630$ and 1200 keV. The main peak, as compared to PWFBA, is closer in width to the experimental one, however approaches it very slowly with the change of the incident proton energy. The use of EWBA (for more details see [13]) does not change the picture for the minimum and for the plateau. Here we need SBA calculations.

3.3. Transfer ionization

In the first series of the TI experiments the scattering plane was fixed by the velocity vectors of the incident proton and the final hydrogen atom. Its azimuthal angle was $\phi = 0$. The cross section was integrated over the perpendicular component of the electron momentum. Integration over the scattering angle was performed within the fixed experimental windows.

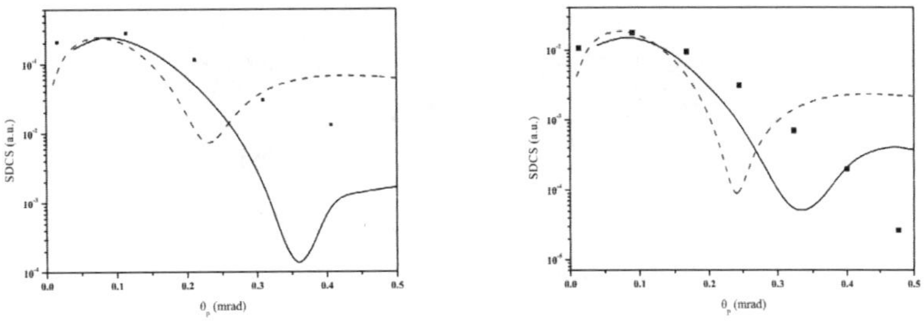

Figure 5. SDCS vs θ_p. The calculation is performed with the eikonal approximation using a 9D code [13] and the SPM wave function. PWFBA is shown with dashed line, EWBA with solid line, experiment [13] with black squares. $n \leq 3$. On the lhs $E_p = 630$ keV, on the rhs $E_p = 1200$ keV.

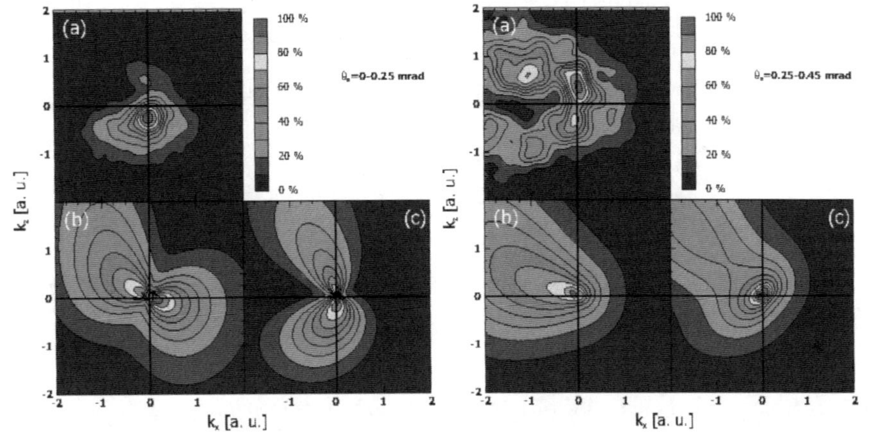

Figure 6. Experimental data (a) and the PWFBA theoretical calculations for TI reactions. The RHF (b, non correlated) and Ch (c, strongly correlated) wave functions are used. $E_p = 630$ keV. On the left panel $0 \leq \theta_p \leq 0.25$, on the right panel $0.25 \leq \theta_p \leq 0.45$. [14]

Triple differential cross section (TDCS) for this kinematical situation is of the form

$$\frac{d^3\sigma}{dk_x dk_z d\phi} = 2\frac{m^2}{(2\pi)^5}\int_{\theta_i}^{\theta_{i+1}} \theta_p d\theta_p \int_{-\infty}^{\infty} dk_y |A_1 + A_2 + A_3|^2. \tag{7}$$

In Fig. 6, left panel, (a) the experimental data for $\theta_p \leq 0.25$ mrad are presented [14]. It is clear that the electron is ejected opposite to the incident proton momentum ($k_z < 0$). It is an indication of the role of the electron-electron correlations. In Fig. 6, left panel, (b) the calculation for the $1s^2$ wave function is shown. For small θ_p the shape of the momentum distribution has the same direct and backward scattering peaks as the $(e, 2e)$ ionization experiments [15]. The comparison of distributions (a) and (b) shows that the helium wave function with no electron-electron correlations does not explain well the experiment. At the same time, the calculation with strong electron-electron correlations, which is presented in Fig. 6, left panel, (c), agrees with experiment much better, but the peak along the incident proton direction $k_z > 0$ is larger.

In Fig. 6, right panel, the data for larger scattering angles are presented. The PWFBA theory now reproduces the experimental data very poorly both for correlated and for non-correlated

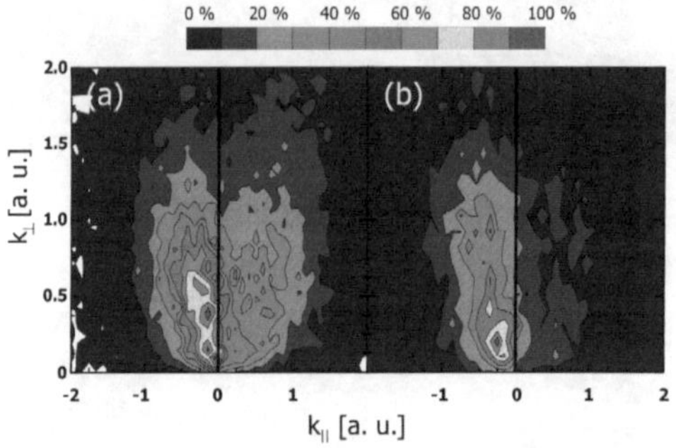

Figure 7. Experimental 2D momentum distribution of the emitted electron from [2]. On the lhs $E_p = 300$ keV, on the rhs $E_p = 630$ keV.

wave functions. One can conclude that the PWFBA describes TI reactions quite well only within the domain of small scattering angles and only in the backward emission direction where the SO mechanism dominates.

The next series of experiments was performed without fixing the scattering plane. The double differential cross section is given by ($k_{||} \equiv k_z$)

$$\frac{d^2\sigma}{dk_\perp dk_{||}} = 2\frac{m_p^2 k_\perp}{(2\pi)^4}\int_0^{\theta_{max}} \theta_p d\theta_p \int_0^{2\pi} d\phi_k |A_1 + A_2 + A_3|^2 \quad (8)$$

The experimental data in Fig. 7 is taken from [2]. Three different helium wave functions are used in calculations: the non correlated RHF wave function and two strongly correlated functions Ch and CI.

One can find in the experimental data shown in Fig. 7 two peaks, the larger one in the backward direction, and the smaller one in the forward direction. Again, the calculation using the wave function with no correlations (see Fig. 8, left panel, (c) and (d)) does not describe the backward-emitted electrons. The calculations with strong correlations exhibit both peaks. To see more clearly the difference between the experiment and calculations, we present the slice of the 2D distributions in Fig. 8, right panel.

In general, PWFBA calculations with a strongly correlated wave function describes the backward-emission peak quite well. However, in the forward direction one should use SBA calculations because the theoretical cross section is much larger than the experimental one. Worth mentioning that both strongly correlated wave functions gave practically the same result. Why using SBA? The PWFBA reflects well the SO mechanism of the electron emission. And this emission is mainly due to the correlations. After colliding with the proton, one electron moves mainly in the forward direction and then it is captured by the proton. The emitted electron is repelled by this fast electron and moves predominantly in the backward direction. The successive mechanism gives another picture, and only a small fraction of it is accounted for by the PWFBA. One can see from the diagram A_2 that a kicked electron is emitted, but it predominantly moves in the forward direction. And most of the SBA diagrams contribute to the forward direction.

4. Conclusions

The CT, TE and TI processes were considered using different Born approximations, mainly PWFBA. The kinematical conditions for the first Born approximation to be valid, namely the energy of the incident proton higher than 500 Kev and small scattering angles near the main

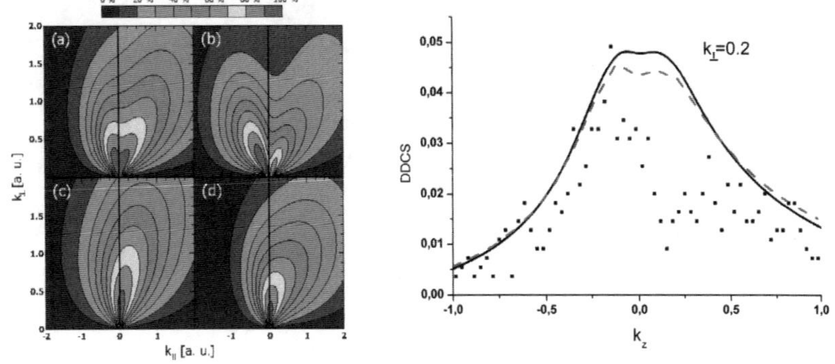

Figure 8. Left panel: theoretical 2D momentum PWFBA distribution of the emitted electron. At the top (a) ($E_p = 300$ keV) and (b) ($E_p = 630$ keV), calculations with the strongly correlated Ch helium initial wave function are presented, at the bottom (c) ($E_p = 300$ keV) and (d) ($E_p = 630$ keV), calculations are performed with the RHF which is non correlated. Right panel: TDCS vs k_z and fixed $k_\perp = 0.2$ a.u., $E_p = 300$ keV. Solid line represents PWFBA with use of the highly correlated wave functions Ch and CI, dots represent the experimental data [14].

peak (fractions of mrad), were determined. If these conditions are fulfilled, even the first Born approximation describes the experiment satisfactorily.

The electron-electron correlations in the initial state play the most important role in the cases of TI and TE reactions. However, to describe the domains of forward electron emission and higher scattering angles, the second Born approximation is needed. Such calculations are in a progress.

Acknowledgments

The authors are grateful to Lomonosov Moscow State University Supercomputing Centre: all calculations were performed using supercomputers Chebyshev and Lomonosov. The work was performed under RFBR grant no. 11-01-00523a. A. G. is grateful to the Dynasty foundation for the financial support.

References

[1] C. D. Lin, J. Phys. B: Atom. Molec. Phys., Vol 11, No. 6, (1978).
[2] M. S. Schöffler *et al.*, Phys. Rev. A, **87**, 032715, (2013).
[3] J. H. McGuire, Electron correlation dynamics in atomic collisions, Cambridge University Press, Cambridge, (1997).
[4] H.-K. Kim *et al.*, Phys. Rev. A, **85**, 022707, (2012).
[5] J.D. Jackson and H. Schiff, Phys. Rev., **89**, 359, (1953).
[6] D. Belkić, I. Mančev, and J. Hanssen, Rev. Mod. Phys. **80**, 249 (2008); D. Belkić, R. Gayet, and A. Salin, Phys. Rep. 56, 279, (1979).
[7] V. L. Shablov et al., Phys. Part. Nucl. **41**, 335, (2010).
[8] O. Chuluunbaatar *et al.*,J. Phys. B: At. Mol. Opt. Phys. ,**74**, 014703, 2006.
[9] J. Mitroy, I. E. McCarthy, and E. Weigold, J. Phys. B **18**, 4149, (1985).
[10] J. N. Silverman, O. Platas, and F. A. Matsen, The journal of Chemical Physics **32**, 1402, (1960).
[11] E. Clementi and C. Roetti, Atomic Data and Nuclear Data Tables **14**, 177, (1977).
[12] S. Houamer, Yu. V. Popov, and C. Dal Cappello, Phys.Lett.A, **373**, 4447, (2009).
[13] M. S. Schöffler *et al.*, Phys. Rev. A **89**, 032707, (2014).
[14] M. S. Schöffler *et al.*, Phys. Rev. A **88**, 042710, (2013).
[15] H. Ehrhardt, K. Jung, G. Knoth, and P. Schlemmer, Z. Phys. D **1**, 3, (1986).

Sturmian approach to single photoionization of many electron atoms and molecules

C M Granados–Castro[1,2], L U Ancarani[1], G Gasaneo[2,4] and D M Mitnik[3,4]

[1] Equipe TMS, SRSMC, UMR CNRS 7565, Université de Lorraine, 57078 Metz, France
[2] Departamento de Física, Universidad Nacional del Sur, 8000 Bahía Blanca, Buenos Aires, Argentina
[3] Instituto de Astronomía y Física del Espacio (IAFE), Universidad de Buenos Aires, Buenos Aires, Argentina
[4] Consejo Nacional de Investigaciones Científicas y Técnicas (CONICET), Argentina

E-mail: carlos-mario.granados-castro@univ-lorraine.fr

Abstract. The application of generalized Sturmian functions to the study of single photoionization of different atomic and molecular systems is illustrated using model potentials. Sturmian functions have been used successfully to study single and double ionization by electron impact on He, while their implementation to molecular systems is still under development. The radial part of the scattering function is expanded in a generalized Sturmian basis. The solution of the resulting matrix problem provides the expansion coefficients. The latter yield directly the photoionization transition amplitude and thus the cross sections. Here we show our preliminary results for He and Ne atoms on one hand, and for CH_4 and H_2O molecules on the other.

1. Introduction

In the past, many different theoretical methods have been developed to study photoionization of atoms and molecules. However, the accurate representation of the continuum states of atomic systems, and more so for molecular systems, still represents a big challenge. In more recent years, the Sturmian approach, using Generalized Sturmian Functions (GSFs), has been applied successfully for the study of several atomic ionization processes [1–4], and the extension of the method to molecular systems is under development.

In the present contribution, we report the preliminary results of the use of the Sturmian approach to study the single photoionization of simple molecules, specially for CH_4 and H_2O. To do so, we first validate our approach by studying many–electron atoms; here we show the cases of He and Ne.

As the description of the electronic structure of molecules is, in general, a difficult task, many different approximations have been developed, such as the one center expansion (OCE) [5, 6], where all the coordinates of electrons are referred to a common center, located in the center of mass of the system. We shall use this approximation, which is known to give reasonable good results for special cases where a central nucleus exists, as for CH_4 and H_2O [5–8].

Atomic units ($\hbar = e = m_e = 1$) are used throughout unless otherwise indicated.

2. Theory

2.1. Photoionization

In order to study single photoionization of atoms and molecules, we start with the non–relativistic time–dependent Schrödinger equation (TDSE), to which a Fourier transform is applied. We use the first order perturbation theory over the wave function of the continuum (final) states, and the dipolar approximation, either length or velocity gauges, to describe the interaction of the radiation field with the system under study. We solve the resulting driven equation, i.e., the time–independent Schrödinger equation (TISE) for the continuum final state wave function $\Phi^{(1)}$

$$\left[\omega_k + \frac{1}{2}\nabla^2 - U(r)\right] \Phi^{(1)}(\mathbf{r}, \omega) = \mathcal{F}(\omega)\, \widehat{D}^{(g)} \Phi^{(0)}(\mathbf{r}), \tag{1}$$

where ω is the energy of the incident photon, $\omega_k = \omega - \omega_0$ is the energy of the resulting photoelectron, $\mathcal{F}(\omega)$ is the Fourier transform of the radiation field, $U(r)$ is the scattering potential felt by the photoelectron, $\Phi^{(0)}(\mathbf{r})$ is the initial state wave function of energy ω_0, and $\widehat{D}^{(g)}$ is the dipolar operator in length (L) or velocity (V) gauges. For the radiation field we consider a laser pulse with a sine squared envelope.

Before solving equation (1), we separate the continuum wave function in its radial and angular parts

$$\Phi^{(1)}(\mathbf{r}, \omega) = \sum_{\ell m} \varphi^{(1)}_{\ell E}(r, \omega)\, Y^m_\ell(\Omega) \tag{2}$$

and the radial part is expanded in a basis set

$$\varphi^{(1)}_{\ell E}(r, \omega) = \frac{1}{r} \sum_n a^{(\ell,E)}_n(\omega)\, S^{(\ell,E)}_n(r). \tag{3}$$

In this work, for the radial function $S^{(\ell,E)}_n(r)$, we choose to use GSFs. Their implementation to study ionization processes in molecular systems has not been done before.

2.2. Generalized Sturmian functions

In general, Sturmian functions are the set of solutions to a Sturm–Liouville problem. In particular, GSFs are solutions (regular at the origin, i.e., $S^{(\ell,E)}_n(r = 0) = 0$), of the radial equation

$$\left(-\frac{1}{2}\frac{d^2}{dr^2} + \frac{\ell(\ell+1)}{2r^2} + \mathcal{U}(r) - E\right) S^{(\ell,E)}_n(r) = -\beta^{(\ell,E)}_n \mathcal{V}(r)\, S^{(\ell,E)}_n(r) \tag{4}$$

where E is an externally fixed parameter, the energy of the continuum state which one wants to describe; $\mathcal{U}(r)$ is called the *auxiliary potential*, usually of long–range, dictating, together with E, the asymptotic behavior of $S^{(\ell,E)}_n(r)$; $\mathcal{V}(r)$ is a short–range *generating potential*; and $\beta^{(\ell,E)}_n$ are the eigenvalues of the problem. Further details on GSFs are given in the review [1].

In this work we use a Yukawa potential with an energy–dependent parameter as generating potential, and a Coulomb potential with charge -1 as auxiliary potential, in order to attribute to the radial functions the correct asymptotic form. The use of GSFs as a basis set allows us to perform very efficient calculations because of their intrinsic physical properties: each function has the exact asymptotic behavior dictated by the auxiliary potential for the given energy E [1]. When photoionizing neutral targets, the continuum final states describing the photoelectron is, asymptotically, a Coulomb wave with charge -1.

2.3. Description of atomic systems

To study single photoionization of atomic systems, we start with the single–active approximation (SAE) [9], where the ionized electron is in a screened potential. For He we use the Herman–Skillman model potential [10]

$$U_{\mathrm{HS}}(r) = \frac{Z_1}{r} + \left(Z_2 + \frac{Z_3}{r}\right) e^{-\alpha r}, \tag{5}$$

with the parameters $Z_1 = Z_3 = -1$, $Z_2 = -2$ and $\alpha = 3.5499$. For Ne we use the model potential given by Rogers *et al* [11]

$$U_{\mathrm{Par}}(r) = -\frac{1}{r}\left[Z - \nu + \sum_{n=1}^{n_{\max}} N_n e^{-\alpha_n r}\right], \tag{6}$$

where ν is the number of electrons in the parental ion, N_n is the number of electrons in the n–th shell, $n_{\max}$ is the maximum value for n in the parental ion, and α_n is a screening parameter.

In figure 1 we plot the effective charges, $rU(r)$, associated with the model potentials (5) and (6) for He and Ne. One sees how the nuclear charge is screened by electrons remaining in the parental ion. The asymptotic charge is always -1, as it should be, indicating that these potentials can be used to study single photoionization.

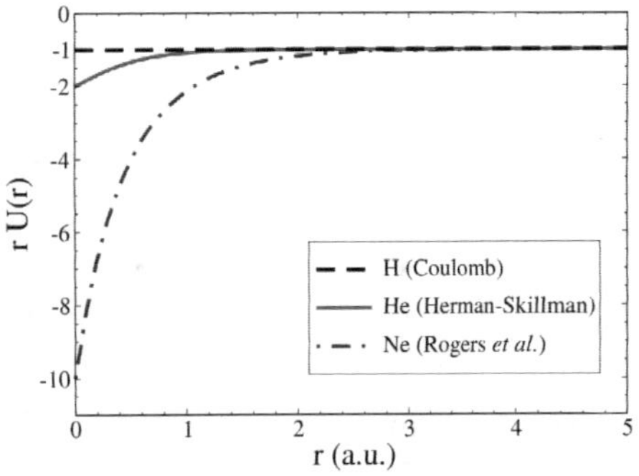

Figure 1. (Color online) Effective charges, $rU(r)$, for H, He and Ne. For H (black – – –) we use the Coulomb potential; for He (red ——) the Herman–Skillman potential (5), and for Ne (blue — · —) the parametric model potential by Rogers *et al* (6).

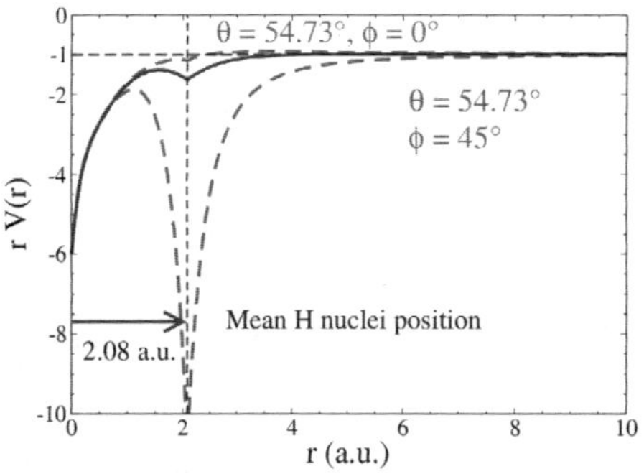

Figure 2. (Color online) Effective charges for CH_4: obtained with the molecular non–central (7) at indicated angles (red and green – – –), and with the angular averaged potential (8) (blue ——). A local minimum is located in the mean position of the H atoms around the C atom, at $R_H = 2.08$ a.u.

The diagonalization of the potential (5) (respectively (6)) provides the ground state description of He (respectively Ne): (*i*) the eigenfunction of the valence electron to be used in the TISE (1); (*ii*) the first ionization energy, reported in table 1.

2.4. Description of molecular systems

In this contribution we use two different model potentials to describe molecular systems. When ionization experiments are performed the real orientation of the molecule is generally undetermined, so that an average over those orientations must be considered. This issue is dealt with an angular average over the molecular potentials before we calculate any continuum

Table 1. Ground state energies ω_0 of valence electrons. For He we use the Herman–Skillman potential (5) and for Ne the parametric potential (6). Each energy is compared with experimental first ionization energies I.

Atom	level	ω_0 (a.u.)	I (a.u.)
He	$1s$	-0.903717	0.903569 [12, 13]
Ne	$2p$	-0.795327	0.792482 [13, 14]

Table 2. Ground state energies ω_0 of valence electrons obtained using the model potential (10), for the studied MOs of CH_4 and H_2O. The energies are compared with the values reported by Moccia [5, 6].

Molecule	MO	level	α	ω_0 (a.u.)	ω_0 (a.u.)
CH_4	$1t_2$	$2p$	-1.381368	-0.504199	-0.5042
H_2O	$1b_1$	$3p$	-0.571085	-0.495400	-0.4954

state. Also, in order to simplify the description of the problem, we use the one–center expansion (OCE) [5]. We start from the molecular model potential given in [15]

$$V_{\mathrm{mol}\,i}(\mathbf{r}) = -\sum_{n=1}^{M} \frac{Z_n}{|\mathbf{r}-\mathbf{R}_n|} + \sum_{j=1}^{N_{\mathrm{MO}}} N_{ij} \int d\mathbf{r}' \frac{|\varphi_j(\mathbf{r}')|^2}{|\mathbf{r}-\mathbf{r}'|}, \tag{7}$$

where M is the number of nuclei, Z_n is the charge of each nuclei, located by $\mathbf{R}_n$, N_{MO} is the number of molecular orbitals (MOs), $N_{ij} = 2 - \delta_{ij}$, and $\varphi_j(\mathbf{r})$ is the wave function of each MO; in this work we use the wave functions given by Moccia in [5] for CH_4 and in [6] for H_2O, both given in terms of Slater–type orbitals. This potential is a non–central potential, and its use in the TISE (1) gives a set of angular–coupled equations. In order to consider the random spatial orientation of molecules, we perform an angular average over it, i.e.,

$$U_{\mathrm{mol}\,i}(r) = \frac{1}{4\pi} \int_{4\pi} d\Omega\, V_{\mathrm{mol}\,i}(\mathbf{r}), \tag{8}$$

yielding a central potential. We plot, for CH_4, the effective charge corresponding to both potentials (7) and (8) in figure 2 where we can observe a local minimum located at the equilibrium distance of the H atoms around the C atom. The depth and width of this minimum varies with the angular orientations (θ, ϕ) of observation.

In order to improve the representation of the molecule given by the angular averaged potential (8), we add to it an exchange potential [16, 17]

$$U_{\mathrm{Exch}\,i}(r) = \alpha\,|\rho_i(r)|^{1/3}, \qquad \rho_i(r) = \frac{1}{4\pi}\left|\frac{\varphi_i(r)}{r}\right|^2, \tag{9}$$

where $\rho_i(r)$ is the electronic density of the initial MO. The parameter α is adjusted as to fix, for each MO, the ground state of the new total potential

$$U'_{\mathrm{mol}\,i}(r) = U_{\mathrm{mol}\,i}(r) + U_{\mathrm{Exch}\,i}(r), \tag{10}$$

to the accurate energy values [5, 6]. In table 2 we show the obtained energies, together with the α values, for the valence MOs of the two studied molecules, CH_4 and H_2O. The corresponding eigenfunctions are used as the initial wave function $\Phi^{(0)}$ in the TISE (1).

3. Results

To study single photoionization in atoms and molecules, we solve the TISE (1), using the potentials (5) or (6) for atomic systems and either (10) or (7) for molecular systems. After expanding the scattering radial wave function (3) in GSFs, the resulting equation is projected over all the elements of the basis set to obtain a linear system of equations for the expansion coefficients. We solve the associated matrix system, and the obtained coefficients can be used to extract the photoionization cross sections, tacking advantage of the GSFs basis asymptotic properties [1]. For all the targets considered in this contribution, we used 60 basis functions.

We show below the results for atoms and molecules separately, and shall compare them with a selection of experimental data and other theoretical calculations (two at maximum). This is enough to show the cross sections general features and the kind of agreement one may find. Although we do not expect the SAE to be sufficiently accurate at low energies, we shall show the cross sections versus photoelectron energy starting from the threshold where theories have most difficulties in reproducing the measurements.

3.1. Atoms

For both He and Ne we calculated the photoionization cross sections from the valence orbital in both length and velocity gauges. For He the results are shown in figure 3. They are compared with an *ab initio* nonresonant cross section by Granados–Castro and Sanz–Vicario [18], and with the experimental data of Samson *et al* [19]; the agreement between our results and other theoretical and experimental data is good, in particular in the length gauge.

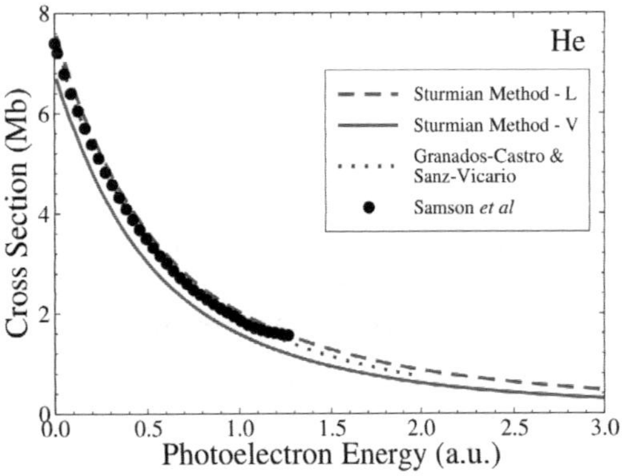

Figure 3. (Color online) Photoionization cross section in Mb vs. photoelectron energy for He. We compare our results calculated in the length (green – – –) and velocity (red ——) gauges with a non–resonant cross section [18] (blue ······) and with experimental data [19] (black ●).

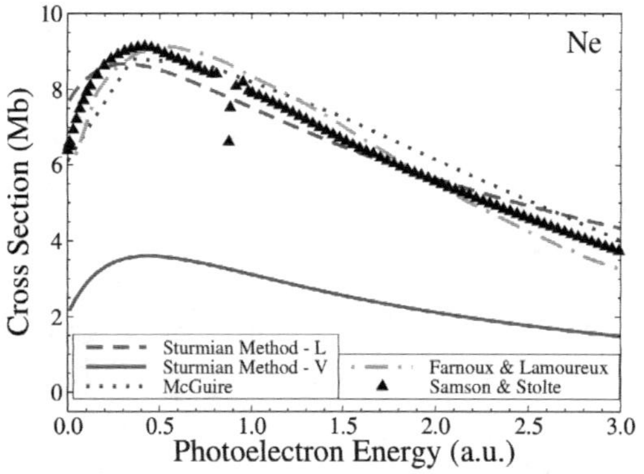

Figure 4. (Color online) Same as figure 3 for Ne. Our results are compared with theoretical results [20] (blue ······) and [21] (orange — · —), and with experimental data [22] (black ▲).

In figure 4 we plot the results for Ne and compare them with the theoretical calculations of McGuire [20] and of Farnoux and Lamoureux [21], and with the experimental data of Samson *et al* [22]. Again our calculations in length gauge show a good agreement. However, a large magnitude difference with the velocity gauge is found. To crosscheck our results, we reproduced them with an equivalent calculation performed using B–splines [23]. We may also exclude any inconsistency or shortcoming of the Sturmian method that could be associated to the $\ell \neq 0$ nature of the initial state orbitals; indeed, the same approach reproduces exactly, within numerical precision, the photoionization cross sections from np orbitals for H. The discrepancy in cross sections obtained with both gauges is known, and has been observed in other calculations for Ne [24] and, for example, in similar approaches for Ar and Xe [25, 26]; it may be explained by a high sensitivity to the quality of the ground state wave function (in particular at short distances from the nucleus), as well the exclusion of different many–body effects as the intrachannel interactions and virtual double excitations in the ground state [26]. It is clearly necessary to use more sophisticated model potentials in order to obtain a better agreement between both gauges. However, the original purpose of this work is to study photoionization in molecules, the atomic case serving to test the methodology and to indicate the range of validity of the SAE.

3.2. Molecules

We now turn to the photoionization from the outer valence MOs of CH_4 and H_2O. The results in length and velocity gauges for the MO $1t_2$ of CH_4 are shown in figure 5. They are compared with the theoretical calculations by Kilcoyne *et al* [27], obtained using the ground state inversion potential method/diffraction, which calculates the total ionization amplitude with a superposition of the atomic ionization amplitudes. The cross section is also compared with the theoretical results by Stener *et al* [8], based on a time–dependent DFT method, and with the experimental data of Backx and van der Wiel [28]. We see that our results obtained using the central potential (10) show an excellent agreement in the asymptotic energy region. For small energies, however, the experimental cross section is largely underestimated. The result of the calculation using the non–central potential (7) in velocity gauge are in a better agreement with the experimental data in all the considered energy range. At small photoelectron energies, the cross section peak is comparable to that obtained by the two other calculations.

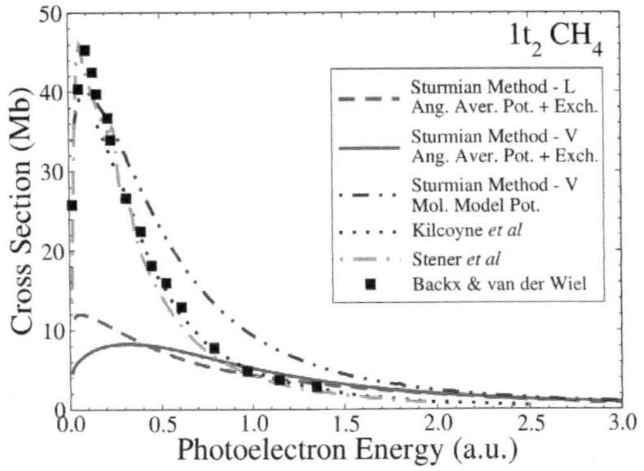

Figure 5. (Color online) Photoionization cross section in Mb vs. photoelectron energy from the MO $1t_2$ of CH_4. We compare our results in length (green – – –) and velocity (red ——) gauges using the central potential (10) and using the non–central potential (7) (blue — · · —), with the theoretical calculations [27] (maroon ······) and [8] (orange — · —), and with experimental data [28] (black ■).

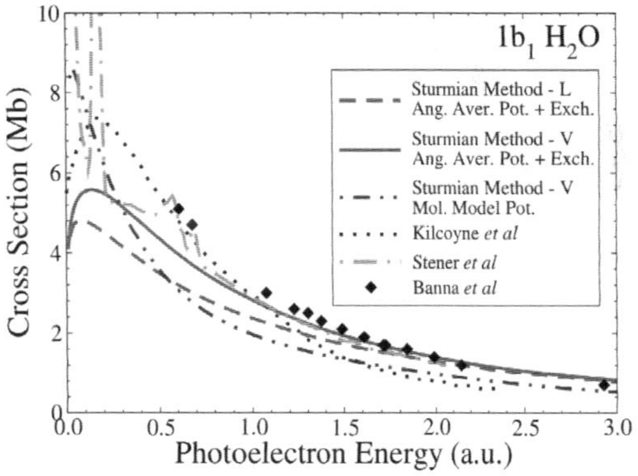

Figure 6. (Color online) Photoionization cross section in Mb vs. photoelectron energy from the MO $1b_1$ of H_2O. We compare our results in length (green – – –) and velocity (red ——) gauges using the central potential (10) and using the non–central potential (7) (blue — · · —), with the theoretical calculations [27] (maroon ······) and [8] (orange — · —), and with experimental data [29] (black ♦).

Our results for the MO $1b_1$ of H_2O are shown in figure 6. They are compared with the results of Kilcoyne *et al* [27] and Stener *et al* [8], and with the experimental data of Banna *et al* [29].

Similar to the results of CH_4, the use of (10) yields a very good agreement in the asymptotic energy region, and this time the region for small energies is slightly better represented. It is worth noting how the three theoretical approaches yield cross sections which differ substantially in the range 0 to 1 a.u. As for CH_4, our results with potential (7) are in reasonable agreement with the experimental data for all the calculated energies.

Similarly to the observation made for atoms, photoionization cross sections for molecular systems are also very sensitive to the selected potential and to the quality of the ground state wave function. This is clearly put in evidence by the different shapes observed in figures 5 or 6. More sophisticated descriptions of both potentials and ground states wave functions will deliver better results, but overall the cross sections presented in these figures have good general features.

4. Summary and perspectives

Considering we are using one–electron wave functions, the GSF method yields good results for the photoionization of atoms, in particular in the length gauge. With a built-in correct asymptotic behavior, the GSF are particularly efficient: using only a relatively small number of basis elements, the method produces converged results (scattering wave functions and transition amplitudes). Turning to applications in molecular systems, the same GSF method combined with the OCE and model molecular potentials, central and non–central, is used to describe the molecular continuum states. Again, using a relatively small number of basis elements, we obtained reasonable results, specially for high photoelectron energies. However, the description of the continuum for lower energies is still incomplete, and a more accurate treatment of the valence MOs and the scattering potentials themselves must be done in order to improve the results. Also full consideration of the spatial orientation of the molecule must be included in our calculations. Work in this direction is on the way.

The next step will be to study single ionization by electron collisions with molecules, using in particular the non–central potential (7). Issues related to the spatial orientation of the molecule are of great importance.

Acknowledgments

We acknowledge the CNRS (PICS project No. 06304) and CONICET (project No. DI 158114) for funding our French-Argentinian collaboration. G. Gasaneo also thanks the support by PGI (24/F059) of the Universidad Nacional del Sur.

References

[1] Gasaneo G, Ancarani L U, Mitnik D M, Randazzo J M, Frapiccini A L and Colavecchia F D 2013 *Adv. Quantum Chem.* **67** 153

[2] Randazzo J M, Buezas F, Frapiccini A L, Colavecchia F D and Gasaneo G 2011 *Phys. Rev. A* **84** 052715

[3] Gasaneo G, Mitnik D M, Randazzo J M, Ancarani L U and Colavecchia F D 2013 *Phys. Rev. A* **87** 042707

[4] Ambrosio M J, Gasaneo G and Colavecchia F D 2014 *Phys. Rev. A* **89** 012713

[5] Moccia R 1964 *J. Chem. Phys.* **40** 2164

[6] Moccia R 1964 *J. Chem. Phys.* **40** 2186

[7] Stener M and Decleva P 1999 *J. Electron Spectrosc. Relat. Phenom.* **104** 135

[8] Stener M, Fronzoni G, Toffoli D and Decleva P 2002 *Chem. Phys.* **282** 337

[9] Fano U and Cooper J 1968 *Rev. Mod. Phys.* **40** 441

[10] McDowell M R C and Coleman J P 1970 *Introduction to the theory of ion-atom collisions* (North-Holland Pub. Co.)

[11] Rogers F J, Wilson B G and Iglesias C A 1988 *Phys. Rev. A* **38**(10) 5007

[12] Kandula D Z, Gohle C, Pinkert T J, Ubachs W and Eikema K S E 2010 *Phys. Rev. Lett.* **105** 063001

[13] Kramida A, Ralchenko Y, Reader J and NIST ASD Team 2014 NIST Atomic Spectra Database (ver. 5.2), [Online]. Available: `http://physics.nist.gov/asd` [2014, July]. National Institute of Standards and Technology, Gaithersburg, MD.

[14] Kaufman V and Minnhagen L 1972 *J. Opt. Soc. Am.* **62** 92
Saloman E B and Sansonetti C J 2004 *J. Phys. Chem. Ref. Data* **33** 1113

[15] Fernández-Menchero L and Otranto S 2010 *Phys. Rev. A* **82** 022712

[16] Hammerling P, Shine W W and Kivel B 1957 *J. Appl. Phys.* **28** 760

[17] Ciappina M F, Pindzola M S and Colgan J 2013 *J. Phys. B: At. Mol. Opt. Phys.* **46** 215206

[18] Granados-Castro C M and Sanz-Vicario J L 2013 *J. Phys. B: At. Mol. Opt. Phys.* **46** 055601

[19] Samson J A R, He Z X, Yin L and Haddad G N 1994 *J. Phys. B: At. Mol. Opt. Phys.* **27** 887

[20] McGuire E 1968 *Phys. Rev.* **175** 20

[21] Farnoux F C and Lamoureux M 1976 *J. Phys. B: At. Mol. Phys.* **9** 897

[22] Samson J and Stolte W 2002 *J. Electron Spectrosc. Relat. Phenom.* **123** 265

[23] Bachau H, Cormier E, Decleva P, Hansen J E and Martín F 2001 *Rep. Prog. Phys.* **64** 1815

[24] Cacelli I, Carravetta V and Moccia R 1983 *J. Phys. B At. Mol. Phys.* **16** 1895

[25] Starace A 1970 *Phys. Rev. A* **2** 118

[26] Starace A F 1983 *Fundamental Processes in Energetic Atomic Collisions* (Plenum Press, NY) p 69

[27] Kilcoyne D A, Nordholm S and Hush N S 1986 *Chem. Phys.* **107** 213

[28] Backx C and Van der Wiel M J 1975 *J. Phys. B: Atom. Molec. Phys.* **8** 3020

[29] Banna M S, McQuaide B H, Malutzki R and Schmidt V 1986 *J. Chem. Phys.* **84** 4739

Target electron ionization in Li^{2+}-Li collisions: A multi-electron perspective

M. D. Śpiewanowski[1], L. Gulyás[2], M. Horbatsch[1], J. Goullon[3], N. Ferreira[3], R. Hubele[3], V. L. B. de Jesus[4], H. Lindenblatt[3], K.Schneider[3], M. Schulz[5], M. Schuricke[3], Z. Song[3], S. Zhang[3], D. Fischer[3], T. Kirchner[1]

[1]Department of Physics and Astronomy, York University, Toronto, Ontario, Canada M3J 1P3
[2]Institute for Nuclear Research, Hungarian Academy of Sciences (ATOMKI), P.O. Box 51, H-4001 Debrecen, Hungary
[3]Max Planck Institute for Nuclear Physics, Saupfercheckweg 1, 69117 Heidelberg, Germany
[4] Instituto Federal de Educação, Ciência e Tecnologia do Rio de Janeiro (IFRJ), Rua Lucio Tavares 1045, 26530-060 Nilópolis/RJ, Brazil
[5]Department of Physics and LAMOR, Missouri University of Science & Technology, Rolla, Missouri 65409, USA

Abstract. Target electron removal in Li^{2+}-Li collisions at 2290 keV/amu is studied experimentally and theoretically for ground and excited lithium target configurations. It is shown that in outer-shell ionization a single-electron process plays the dominant part. However, the K-shell ionization results are more difficult to interpret. According to our calculations, the process is shown to be strongly single-particle like. On one hand, a high resemblance between theoretical single-particle ionization and exclusive inner-shell ionization is demonstrated, and contributions from multi-electron processes are found to be weak. On the other hand, it is indicated by the discrepancy between experimental and single-particle theoretical results that multi-electron processes involving ionization from the outer-shell may play a crucial role.

1. Introduction

Electron removal from few-electron atoms by fast ions has been of interest to collision physics for many decades [1]. The great interest for these collision systems is mostly fueled by a desire for a better understanding of single- and multi-electron processes. Nevertheless, technical difficulties led to an emphasis on helium atoms among a number of possible simple few-electron atomic targets. Only recently, with the experimental development of the magneto-optical trap reaction-microscope (MOTReMi) setup [2] did it become possible to investigate a whole new range of collision systems allowing to study single- and multi-electron processes with great accuracy. The MOTReMi approach is suitable for alkalis, and lithium has become experimentally accessible as a target atom [2], raising substantial experimental and theoretical interest [3, 4, 5, 6, 7].

In this paper we investigate target ionization in Li^{2+}-Li collisions at 2290 keV/amu for $Li(1s^22s)$ (ground) and $Li(1s^22p)$ (excited) target configurations. We present measurements of single-differential cross sections (SDCS) and investigate single- and multi-particle processes that contribute to $1s$, $2s$, and $2p$ vacancy production. The paper is organized as follows. In Sec. 2.1 we introduce details of the MOTReMi experiment. Theoretical details on the solution of the time-dependent Schrödinger equation (TDSE) with various approximations and how to incorporate multi-electron effects into a single-particle model are briefly discussed in Sec. 2.2. Section 3 provides experimental and theoretical results and their discussion. Conclusions are presented in Sec. 4.

2. Methods

2.1. Experimental methods

The experiment was performed using a MOTReMi [2, 8] in the ion storage ring TSR at the Max Planck Institute for Nuclear Physics in Heidelberg. Li^{2+} ions were accelerated by a tandem accelerator and stored in the TSR in bunches of about 2 ns duration. The lithium target cloud was laser-cooled to about 2 mK in a MOT. Recoil ions and electrons produced in ionizing collisions were extracted by electric and magnetic fields towards position and time sensitive detectors in a ReMi [9]. While for the recoil ions the final momenta were typically much smaller than the spectrometer acceptance, electrons could only be collected for transverse momenta smaller than 1.9 a.u.

Due to the interaction of the atoms with the cooling lasers, a certain fraction of the target atoms (approximately 20%) is initially in an excited $1s^22p$ configuration. This made it possible to simultaneously measure cross sections for ionization from the ground state and from the excited state configuration. This is achieved by periodically switching the cooling lasers on and off. During the period with the lasers being switched off 100% of the target state population is in the ground state. During the remaining time a fraction of about 20% of the atoms is in the excited $1s^22p$ state. Cross sections for the ionization from the excited state are obtained by subtracting the spectra acquired during the two different time periods from each other [3, 8]. It should be noted, that the Zeeman splitting of the atoms in the spectrometer magnetic field (whose orientation nearly coincides with the projectile beam direction) results in the polarization of the $2p$ state and about 85% of the excited atoms populate the magnetic sub-level with $m_L = -1$ [8].

Vacancy production in the target K-shell was distinguished from valence vacancy production due to different energy losses $E_{\mathrm{p}}^{\mathrm{loss}}$ of the projectile for both channels. $E_{\mathrm{p}}^{\mathrm{loss}}$, in turn, is given by $(p_{elz} + p_{recz})v_p$ [9], where p_{elz} and p_{recz} are the longitudinal electron and recoil-ion momentum components, respectively, and v_p is the projectile speed. In the present experiment, the longitudinal electron- and recoil-ion momentum resolutions translate into an energy loss resolution of 30eV (FWHM) which is significantly smaller than the difference between the energy losses for K-shell and valence vacancy production (approximately 65 eV).

2.2. Theoretical methods

Our objective is to solve the Schrödinger equation for Li^{2+}-Li collisions with the target in ground and excited configurations. Since the full solution to this multi-electron problem is beyond the reach of the state-of-the-art methods used in collision physics, we apply a series of approximations. First, we employ the semi-classical approximation (SCA), where we assume that the projectile is treated as a particle moving along a classical trajectory. This leaves us with a many-electron TDSE in order to deal with the electronic degrees of freedom. Secondly, the projectile electron is frozen, hence, its only influence on the process is assumed to be an additional potential, which screens the nuclear charge and is obtained from a $1s$-like charge distribution. In a third step, we reduce the Hamiltonian to a set of one-body operators neglecting the full account of electron-electron interactions, i.e., the independent electron (IEL) model is employed. The effective potential that represents the interactions in the target Li atom is approximated by the exchange-only optimized potential method (OPM) of density functional theory [10]. For more details see [7]. We note that the same potential is used to define the ground and excited states of the target atom. No significant changes are expected by taking the proper effective potential for the excited-state lithium atom due to the tight binding of the inner electrons.

Table 1. Single-particle total cross sections for He^{2+}-Li and Li^{2+}-Li collisions at 2290 keV/amu for $1s$, $2s$, and $2p$ initial states relevant for ground and excited target configurations (in units of a_0^2, where a_0 is the Bohr radius).

Projectile	Method	Li(1s)	Li(2s)	Li($2p_{m=0}$)	Li($2p_{\|m\|=1}$)
He^{2+}	CDW-EIS	0.32	2.08	4.33	4.93
	TC-BGM	0.35	2.17	5.58	6.01
Li^{2+}	TC-BGM	0.47	2.30	6.06	6.52

Single-particle TDSEs are solved using the two-center basis generator method (TC-BGM) with a basis that consists of $1s - 4f$ target states, and $1s - 4f$ projectile states together with 71 BGM pseudostates [11]. The projectile states are obtained by solving the time independent Schrödinger equation with the following Hamiltonian (in atomic units, $\hbar = m_e = e = 1$)

$$H = -\frac{1}{2}\Delta - \frac{Z}{r} + \frac{1}{r}\left(1 - (1 + Zr)e^{-2Zr}\right), \tag{1}$$

where r denotes the distance between the projectile nucleus and the (passive) electron and $Z = 3$. The last term originates from freezing the projectile electron in the $1s$ state. All basis states are endowed with electron translation factors to ensure Galilean invariance.

Within the single-active electron (SAE) model one simply takes the single-particle solutions for a given initial state (multiplied by two in the case of the doubly occupied $1s$ orbital) as the final results. To include multi-electron phenomena into the method consistently with the IEL model, we turn to a determinantal analysis [12]. In short, the solutions to the single-particle TDSEs are assembled to form a Slater determinant which is projected onto the final-state determinants. The transition probabilities of interest are thus obtained from combinations of determinants constructed from one-particle density matrix elements, as described in [7].

Extracting electron-emission-energy differential information from TC-BGM solutions is a difficult task. Therefore, we calculate continuum distorted-wave eikonal initial-state (CDW-EIS) single-electron probabilities for inner- and outer-shell ionization using the same target Hamiltonian as in the TC-BGM calculations. Moreover, we restrict our CDW-EIS calculations to the He^{2+}-Li collision system, neglecting the influence of the residual electron in the Li^{2+} projectile. Hence, our SDCSs for multi-electron processes are calculated by combining TC-BGM probabilities for excitations with doubly-differential cross sections obtained from the CDW-EIS model.

The validity of this approach is confirmed by table 1, where the single-particle total cross sections (TCS) for He^{2+}-Li collisions obtained with CDW-EIS are compared with those for He^{2+}-Li and Li^{2+}-Li collisions obtained with TC-BGM. The single-particle TCSs for ionization in He^{2+}-Li collisions for both methods are in good agreement, the differences are within 10% for both inner- and outer-shell ionization of the ground state. The agreement for outer-shell ionization of the excited-state target is not that good and the difference is around 20%. As expected, the situation is slightly worse when comparing He^{2+}-Li collision results with Li^{2+}-Li data, due to the additional interaction that originates form the presence of the projectile electron.

To understand the origin of the discrepancies between different single-particle TCSs in table 1, we compare in figure 1 ionization probabilities for these collision systems obtained from the CDW-EIS and TC-BGM approaches. The $1s$ ionization probability curves for He^{2+}-Li collisions obtained with both methods fall almost on top of each other (see figure 1 (a)) yielding similar

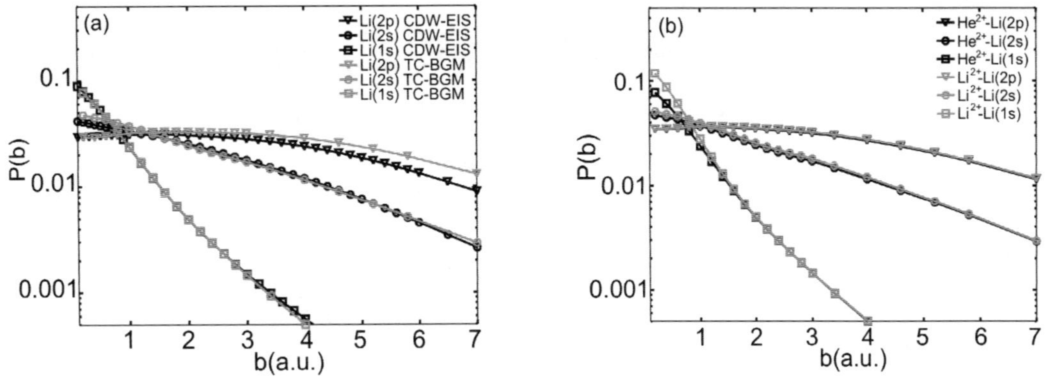

Figure 1. Inner- and outer-shell ionization probabilities in He^{2+}-Li($1s^2 2s$), He^{2+}-Li($1s^2 2p$), Li^{2+}-Li($1s^2 2s$), and Li^{2+}-Li($1s^2 2p$) collisions at 2290 keV/amu. (a) Comparison of TC-BGM and CDW-EIS results for He^{2+}-Li($1s^2 2s$) and He^{2+}-Li($1s^2 2p$) collisions. (b) Comparison of TC-BGM results for Li^{2+}-Li($1s^2 2s$) and Li^{2+}-Li($1s^2 2p$) collisions and TC-BGM results for He^{2+}-Li($1s^2 2s$) and He^{2+}-Li($1s^2 2p$) collisions.

single-particle TCSs. A small difference in ionization from $2s$ can be seen for low values of the impact parameter. It is also clear from the figure that the CDW-EIS method underestimates the ionization probability from the $2p$ state for the whole impact-parameter range, as compared to TC-BGM results. For the $2p$ state we assumed an m level distribution in accord with the experiment, i.e., $|m| = 1$ is populated with 85% and $m = 0$ with 15%. This procedure is adopted for the results that follow.

Figure 1 (b) depicts TC-BGM ionization probabilities for He^{2+}-Li and Li^{2+}-Li collisions. The change in the projectile potential yields changes in ionization probabilities. The potentials for both projectiles have the same Coulomb tail, and it is only in the vicinity of the projectile (less than ~2 a.u.) that the projectile potentials differ significantly. Therefore, the biggest changes in the probabilities ought to be observed for low values of the impact parameter. The change is more pronounced for $1s$ ionization, due to the fact that the $1s$ wave function is compact around the atomic center and is without a node near or at the center, unlike the $2s$ and $2p$ wave functions, respectively. As a result, it is inner-shell ionization that changes the most in comparison with He^{2+}-Li collisions.

Since collisions with small impact parameters are most likely to produce electrons with energies too high to be detected with the apparatus used in this work, we disregard the differences in projectile (Li^{2+} vs He^{2+}) as irrelevant to our work and use the CDW-EIS method for He^{2+}-Li collisions henceforth to describe the Li^{2+}-Li experiments. Finally, we note that according to our TC-BGM calculations the probability for electron capture by the projectile is negligibly small, therefore it is ignored in our analysis.

3. Results

Figure 2 presents experimental and theoretical SDCSs for ionization from the ground and excited states of the lithium target. The theoretical SAE model SDCSs for ionization of valence electrons (triangles), for both ground and excited Li targets, are in a very good agreement with the experimental data (diamonds). The removal of the valence electron for this system can, therefore, be understood as a pure single-particle process. The perturbation is too weak for the tightly bound inner-shell electrons to participate actively in the removal of the loosely bound valence

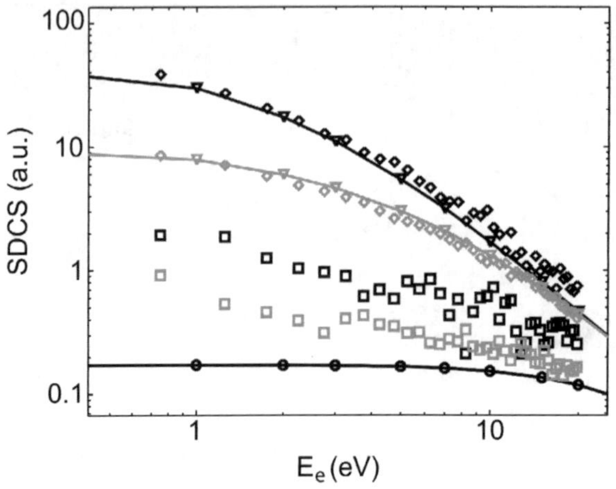

Figure 2. SDCSs for electron removal in Li^{2+}-Li($1s^22s$) (orange) and Li^{2+}-Li($1s^22p$) (black) collisions. Diamonds: experimental SDCSs for outer-shell ionization, squares: experimental SDCSs for inner-shell ionization, triangles: SAE model SDCSs for outer-shell ionization, circles: SAE model SDCSs for inner-shell ionization, which does not distinguish between the two Li-target configurations. The theoretical results were connected by straight lines.

electron.

In the case of Li($1s$) ionization we observe a big discrepancy between the SAE model CDW-EIS (circles) and the experimental (squares) results, reaching over an order of magnitude for the lowest-energy electrons removed from the excited-state Li target. This shows that the ionization of inner-shell electrons is a more complex process. To investigate this difference we follow the work presented in [7] and incorporate many-body phenomena in a way briefly discussed in the previous section.

There are many possible paths that can lead to $1s$ vacancy production leaving the target in an ionic excited configuration. Among the most important are: (i) Exclusive ionization, where one $1s$ electron is removed, whereas the two electrons that are left behind remain unaffected; (ii) Excitation-ionization, with one-electron removal from the inner shell, one-electron excitation from the outer shell, and the residual inner-shell electron remaining unaffected; (iii) Excitation-ionization, with one-electron removal from the outer shell, one-electron excitation from the inner shell, and the residual inner-shell electron remaining unaffected; (iv) Shake-off, here one $1s$ electron is excited while the $2s$ electron is shaken-off ending up in the continuum, whereas the residual inner-shell electron remains unaffected.

Moreover, we consider process (iii) as a sequential process in an independent event (IEV) model where firstly, the valence electron is removed and only after the subsequent rearrangement of the inner-shell electrons, a single $1s$ electron gets promoted to an excited state of the lithium ion. In order to calculate probabilities for elastic scattering and excitation of a $1s$ electron in Li^+ we replace the OPM potential for Li($1s^22s$) with the OPM potential for Li^+($1s^2$). Adjustment of the $2s$ electron to the excited ionic state after inner-shell ionization (process (ii)) appears to be much less likely to happen. Hence, we restrict ourselves to consider only process (iii) within the IEV model.

Our last amendment to the model addresses the restricted target basis used in the TC-BGM calculations. The basis set used here allows excitations to states with principal quantum number not higher than $n = 4$. In our analysis we extend the number of possible excitations to exceed

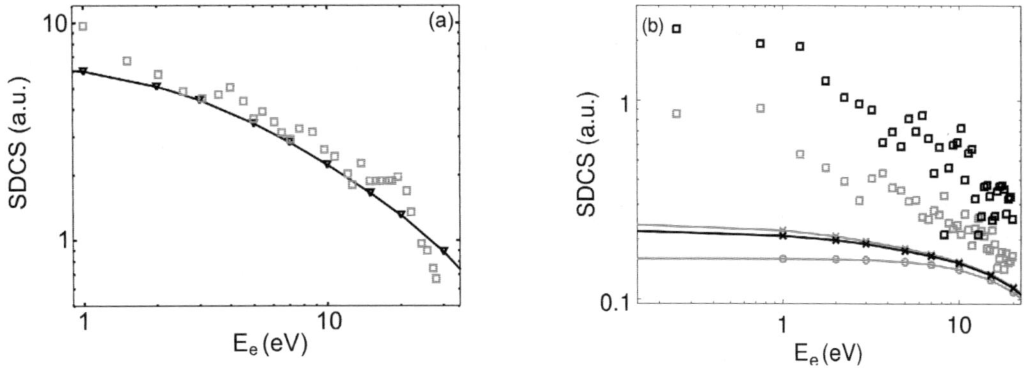

Figure 3. (a) O^{8+}-Li($1s^22s$) SDCS for inner-shell ionization. Theory - Triangles connected by full line (from [7]), experiment - squares (from [2]). (b) SDCSs for the inner-shell electron removal in Li^{2+}-Li($1s^22s$) (orange) and Li^{2+}-Li($1s^22p$) (black) collisions. Squares depict experimental SDCSs. Crosses depict SDCSs that take into account processes (i)-(iv), and circles depict SAE model SDCSs.

$n = 4$ in an approximate way, by assuming that they scale like $1/n^3$ [13]. This allows us to extrapolate excitation probabilities for principal quantum number $n \to \infty$. The effect of the extrapolation is to provide a 10% to 50% increase in SDCSs.

Taking into account the aforementioned paths, with a leading role of the excitation-ionization process, together with incorporating the IEV model, and extrapolation of excitation probabilities, allowed Kirchner and co-workers [7] to explain successfully the experimental SDCS for $1s$ ionization in O^{8+}-Li($1s^22s$) collisions at 1500 keV/amu (see figure 3 (a)). We perform a similar analysis to understand the experimental SCDSs for removal of a $1s$ electron in figure 2.

Figure 3 (b) depicts experimental SDCSs for removal of a $1s$ electron (squares) by Li^{2+} impact, together with theoretical SDCSs where processes from (i)-(iv) are accounted for (crosses) and with SAE model SDCSs (circles). It can be clearly seen that for both ground and excited states of the Li target these multi-electron processes do not contribute sufficiently to explain the experimental results. Another interesting feature is that the ground-state SDCS, in the low-energy part, has a greater magnitude than the excited-state SDCS, contrary to the experimental results. Note, that the SAE model curves fall on top of each other, since we use the same effective potential representing the interactions of ground- and excited-state configurations.

We now turn our attention to each process from (i) to (iv) separately, and to their contributions to the total multi-electron SDCSs (squares in figure 4). The exclusive ionization process (triangles in figure 4) dominates the total SDCSs for inner-shell electron removal for both ground and excited states of the target. Moreover, the resemblance with the SAE model SDCSs (see figure 3 (b)) is striking. This means that the probabilities for leaving both inner- and outer-shell electrons unaffected are close to unity, ruling out any other contribution as significant. The other contributions are indeed considerably weaker and do not constitute even 40% of the total SDCSs. These contributions are not sufficient to explain the discrepancy between theory and experiment. Note that an extra contribution to the collision with excited lithium atoms has been added (dashed line in figure 4), namely deexcitation-ionization, where the outer-shell electron gets deexcited ($2p \to 2s$) while one of the inner-shell electrons is ionized.

Although the discrepancy between experimental and theoretical data is not fully understood we may speculate that a different multi-electron process (or processes) may be involved. In

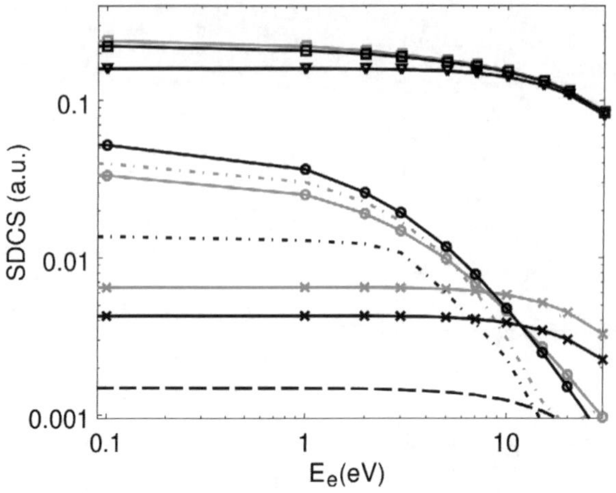

Figure 4. Multi-electron contributions to the total SDCSs for Li($1s^22s$) (orange curves) and Li($1s^22p$) (black curves) target configurations. Triangles: process (i), crosses: process (ii), circles: process (iii) in the IEV model, dashed-dotted: process (iv), dashed: deexcitation-ionization (see the text), squares: total SDCS.

particular, it could be a process that involves outer-shell ionization, as can be seen from the slopes of the experimental SDCS results. The outer-shell-ionization SDCS decreases much more rapidly with electron energy than inner-shell-ionization SDCS. Since the deficiency in the theoretical result lies in the low-energy part, it is a process with outer-shell ionization involvement that could reduce this difference, provided it is of sufficient magnitude.

One candidate is a direct interaction of the projectile electron with the valence target electron. It was shown that such an interaction may lead to simultaneous target ionization and projectile excitation or ionization [14, 15]. Theoretical estimates of these processes result in at most small contributions to the SDCS. Nevertheless, it would be interesting to look for experimental signatures of projectile electron transitions. For example, a measurement of the final projectile charge state could be used to rule out projectile ionization.

4. Conclusions

We have shown that due to the strong binding of the inner-shell electrons, outer-shell ionization in Li^{2+}-Li($1s^22s$) and Li^{2+}-Li($1s^22p$) collisions at 2290 keV/amu is solely a one-electron process. This was also observed for collision systems with a much stronger perturbation [2].

The theoretical results for $1s$ electron removal do not reproduce the experimental results. Moreover, the comparison has brought us to contradictory conclusions. First, the dominant contribution of $1s$ exclusive ionization to the total SDCS and its resemblance to the SAE model SDCS suggest that the process has, at least to a high degree, a single-electron nature. Secondly, the discrepancy between the full multi-electron SDCS and the experiment indicates that a significant multi-electron process or processes involving ionization from the valence shell are missing from our analysis.

Acknowledgements
The theoretical portion of this work was supported by the Natural Sciences and Engineering Research Council of Canada and by the Hungarian Scientific Research Fund (OTKA Grant No.

K 109440). The experiments were funded through the Emmy-Noether program of the German Research Council (DFG), under Grant No. FI 1593/1-1 and were supported by the Alliance Program of the Helmholtz Association under Grant No. HA216/EMMI. MS acknowledges support from the National Science Foundation (Grant No. PHY-1401586).

References

[1] Schulz M and Madison D H 2006 *Int. J. Mod. Phys. A* **21** 3649

[2] Fischer D, Globig D, Goullon J, Grieser M, Hubele R, de Jesus V L B, Kelkar A, LaForge A, Lindenblatt H, Misra D, Najjari B, Schneider K, Schulz M, Sell M and Wang X 2012 *Phys. Rev. Lett.* **109**(11) 113202

[3] LaForge A C, Hubele R, Goullon J, Wang X, Schneider K, de Jesus V L B, Najjari B, Voitkiv A B, Grieser M, Schulz M and Fischer D 2013 *J. Phys. B* **46** 031001

[4] Hubele R, LaForge A, Schulz M, Goullon J, Wang X, Najjari B, Ferreira N, Grieser M, de Jesus V L B, Moshammer R, Schneider K, Voitkiv A B and Fischer D 2013 *Phys. Rev. Lett.* **110**(13) 133201

[5] Ciappina M F, Pindzola M S and Colgan J 2013 *Phys. Rev. A* **87**(4) 042706

[6] Walters H R J and Whelan C T 2014 *Phys. Rev. A* **89**(3) 032709

[7] Kirchner T, Khazai N and Gulyás L 2014 *Phys. Rev. A* **89**(6) 062702

[8] Hubele R, Schuricke M, Goullon J, Lindenblatt H, Ferreira N, Laforge A, Brühl E, de Jesus V L B, Globig D, Kelkar A, Misra D, Schneider K, Schulz M, Sell M, Song Z, Wang X, Zhang S and Fischer D 2014 (*Preprint* arXiv:1411.0838)

[9] Ullrich J, Moshammer R, Dorn A, Dörner R, Schmidt L P H and Schmidt-Böcking H 2003 *Rep. Prog. Phys.* **66** 1463

[10] Engel E and Vosko S H 1993 *Phys. Rev. A* **47**(4) 2800–2811

[11] Zapukhlyak M, Kirchner T, Lüdde H J, Knoop S, Morgenstern R and Hoekstra R 2005 *J. Phys. B* **38** 2353

[12] Lüdde H J and Dreizler R M 1985 *J. Phys. B* **18** 107

[13] Röhrbein D, Kirchner T and Fritzsche S 2010 *Phys. Rev. A* **81**(4) 042701

[14] Dörner R, Mergel V, Ali R, Buck U, Cocke C L, Froschauer K, Jagutzki O, Lencinas S, Meyerhof W E, Nüttgens S, Olson R E, Schmidt-Böcking H, Spielberger L, Tőkesi K, Ullrich J, Unverzagt M and Wu W 1994 *Phys. Rev. Lett.* **72**(20) 3166

[15] Zhang S F, Fischer D, Schulz M, Voitkiv A B, Senftleben A, Dorn A, Ullrich J, Ma X and Moshammer R 2014 *Phys. Rev. Lett.* **112**(2) 023201

Spin-polarized single- and two-electron spectroscopies at various kinematics

S Samarin[1,2], M Tan[1], O Artamonov[2], A Baraban[2], P Guagliardo[3] and J Williams[1]

[1]Centre for Atomic, Molecular and Surface Physics, School of Physics, UWA, Australia
[2] Fock Institute of Physics, St Petersburg State University, Russia
[3]Centre for Microscopy, Characterisation and Analysis, UWA, Perth, Australia

E-mail: samar@physics.uwa.edu.au

Abstract. We present experimental results on the application of spin-polarized single- and two-electron spectroscopy in various kinematics for studying W(110) crystal and ferromagnetic films of Co and Fe. Spin effects in the measured spectra are expressed in terms of the asymmetry A defined as $A=(I^+ - I^-)/(I^+ + I^-)$, where I^+ and I^- are spectra measured for two opposite orientations of the polarization vector of the incident beam. It is demonstrated that the asymmetry spectra depend on the kinematics of experiment and can even change the sign of the asymmetry.

1. Introduction

Spin-polarized Electron Energy Loss Spectroscopy (SPEELS) was applied for studying Stoner excitations [1-8] as well as spin waves on various ferromagnetic surfaces [9, 10]. Spin-polarized two-electron spectroscopy (SPe,2e) provided information on spin-related properties of nonmagnetic [11] and magnetic [12] surfaces. Usually the geometrical arrangement of a SPEELS measurement is characterized by a fixed angle between the incident electron beam and the axis of the detector (analyzer) [7] and the sample can be rotated around the axis perpendicular to the scattering plane containing the electron beam and the detector. In that way the angles of incidence and detection can be varied conveniently.

There are several important specific features in electron scattering from surfaces:

- The presence of the surface implies that the parallel-to-the-surface electron momentum is conserved;
- Refraction of the electron trajectories at the surface potential barrier while passing through the interface;
- Ordered distribution of scattering centres at the surface implies that electrons (or electron pairs) undergo a diffraction;
- Imbalance of spin-up and spin-down electrons in the valence band in a ferromagnetic surface leads to the intensity asymmetry of measured spectra.

Generally two regimes of the SPEELS measurements are used: i) specular geometry, when the incidence angle is equal to the detection angle; and ii) off-specular geometry when the incident angle is not equal to the detection angle. It is assumed that in the specular geometry the mechanism of the electron energy loss is of dipole type, i.e. the energy loss does not involve a large momentum transfer and the interaction of the incident electron with the surface occurs through the electromagnetic field generated by the electron. In the second case, off-specular geometry, the electron-electron encounter occurs with substantial momentum transfer from the incident electron to the valence electron of the target. It is also assumed that in the dipole-type of scattering no exchange effects can be observed.

The aim of this work is to test various kinematical arrangements for the spin-polarized electron spectroscopies and to study the influence of the kinematics on spin effect observed in SPEELS and (SPe,2e).

2. Experimental details

In our experimental set-up we used a spin-polarized electron source based on the photoemission from strained GaAs crystal [13] activated by successive Cs deposition and oxygen adsorption to obtain a negative electron affinity surface. Electrons emitted from this surface under properly tuned laser light are longitudinally polarized with polarization vector either parallel or anti-parallel to the electron momentum depending on the helicity of the laser light. Bending electron trajectories by 90^0 using a quasi-spherical deflector transforms the polarization of the beam from longitudinal to transverse. The sense of electron beam polarization ("up" or "down") can be changed by changing the helicity of the laser light used for extracting electrons from the photocathode.

We used a time-of-flight (TOF) technique for the electron energy measurements [14] for which the electron beam was pulsed with a pulse width of about 1 ns and the repetition rate of 2.5 MHz. The pulsing provides reference point on the time scale. The energy resolution of the TOF method varies with the electron energy and is usually better for low-energy electrons. We estimate the energy resolution of the TOF spectrometer is on average better than 0.5 eV in the energy range of the measurements.

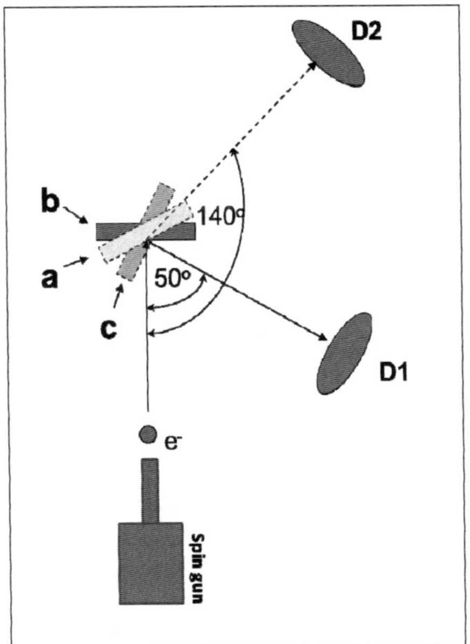

Figure 1. Three possible geometrical configurations of experiment.

A particular feature of our measurements is the possibility of using various kinematics in a wide range of angles of incidence and detection. Figure1 represents a sketch of the geometrical arrangement with two detectors situated at 50^0 and 140^0 with respect to the axis of the spin-polarized electron gun. The sample was mounted on a rotatable manipulator with axis perpendicular to the scattering plane enabling a polar angle change. In addition, the sample can be rotated around the axis perpendicular to its surface for azimuthal angle changes. Three possible polar orientations of the sample are shown in figure1. Position a) and position b) correspond to a specular and normal incidence geometry, respectively, for detection of scattered electrons by Detector 1. Position c) allows the detection of scattered electrons by detector 1 and detector 2 simultaneously. They can be used independently (two SPEELS spectra) or in coincidence (spin-polarized (e,2e) spectroscopy [15, 16, 17]).

As samples we used a single crystal of W(110) and thin ferromagnetic films (Fe, and Co) deposited on the W(110) substrate. Prior to the measurements on W(110) or a deposition of the film, the W(110) crystal surface was cleaned using well established procedure [18, 19] including oxygen adsorption at the sample temperature about 1600 K with subsequent high temperature flashes up to 2500 K.

3. Results and discussion

3.1. Single crystal of tungsten W(110)

The low index faces of a tungsten single crystal, (100), (110) and (111) are very well studied by various electron spectroscopies including Electron Energy Loss Spectroscopy [20, 21], photoelectron spectroscopy [22], secondary emission spectroscopy [23], and two-electron spectroscopy [24]. The W(110) surface is non-reconstructing and very often is used as a substrate for metal films deposition. The large atomic number metal tungsten crystal is a suitable sample for the observation of spin-orbit effects as observed in W(110) using spin-polarized secondary emission spectroscopy [23] and spin-polarized low-energy two-electron spectroscopy [24]. Our aim here is to show how the kinematics of the measurement influences the manifestation of spin-orbit interaction.

Since the spin-orbit effect is proportional to the scalar product of the electron orbital momentum and the electron spin (**L**·**S**), it is a maximum when the polarization vector of the incident electron is perpendicular to the scattering plane (because the orbital angular momentum of the electron is perpendicular to the scattering plane). In all cases the polarization vector was perpendicular to the scattering plane. We used various geometric arrangements to measure the asymmetry A of the scattered and emitted electrons with the asymmetry defined as $A = (I^+ - I^-)/(I^+ + I^-)$, where I^+ and I^- are the energy distributions of scattered electrons for two opposite orientations of the polarization vector of the incident beam.

The presence of the surface breaks the translational symmetry in one direction, perpendicular to the surface, and, as a consequence, only the parallel-to-the-surface component of the electron momentum is conserved (modulo a reciprocal lattice vector). The crystal structure of the sample implies that a reciprocal lattice vector **K** can be added or subtracted in the momentum balance of the electron-electron collision. Figure 2 shows the balance of energy and momenta in simple electron-electron scattering.

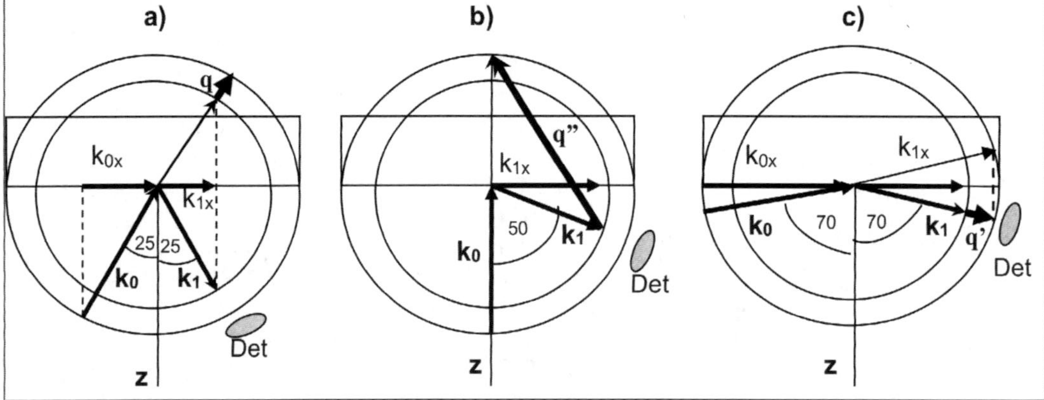

Figure 2. Geometrical representation of the kinematics of electron-electron scattering: $\mathbf{k_0}$ and $\mathbf{k_1}$ are the wave vectors of incident and scattered electrons; k_{0x} and k_{1x} are projections of the incident and scattered electrons on the sample surface; **z** is the axis perpendicular to the sample surface.

In figure 2, sections a), b) and c) correspond to the geometrical arrangements a), b) and c) of figure1, respectively. Wave vectors of the incident and scattered electrons are denoted by $\mathbf{k_0}$ and $\mathbf{k_1}$, respectively. The k_{0x} and k_{1x} are the projections of the incident and scattered electrons. Given the periodicity along the sample surface, the parallel to the surface component of the electron wave vector should satisfy the following equation: $k_{0x} - k_{1x} = q_x \pm nK_x$, where $\hbar\mathbf{q}$ is the momentum transfer, n is integer, **K** is the two-dimensional reciprocal lattice vector. In the case of a creation of an electron-hole pair the momentum of the electron-hole pair is $\hbar\mathbf{q}$. The kinematics presented in figure 2 do not account for the second electron that remains inside the solid. In some cases the electron-electron scattering may be assisted by the reciprocal lattice vector transfer (*umklapp* process).

An example of a SPEELS measurement on W(110) crystal at three different kinematics is shown in figure 3. Low energy electrons (13 eV) impinge onto the surface at normal incidence and secondary electrons (including true secondaries, elastically and inelastically scattered primary electrons) are detected by the right (Det. 2) or the left (Det. 1) detectors as shown on the section a) of figure 3. The corresponding asymmetries A, red circles (Det. 2) and black squares (Det. 1) are mirror symmetric with respect to the x-axis. This symmetry is consistent with the general symmetry of the experimental set-up; specifically the sample crystal is symmetric with respect to the plane perpendicular to the sample surface and to the scattering plane. Mirror reflection in this plane of the whole experimental set-up will exchange detectors Det. 1 and Det. 2 and flip the polarization vector of the incident beam (because it is an axial vector) with everything else remaining unchanged. Therefore the asymmetries measured in Det. 1 and Det. 2 are the same but with opposite signs.

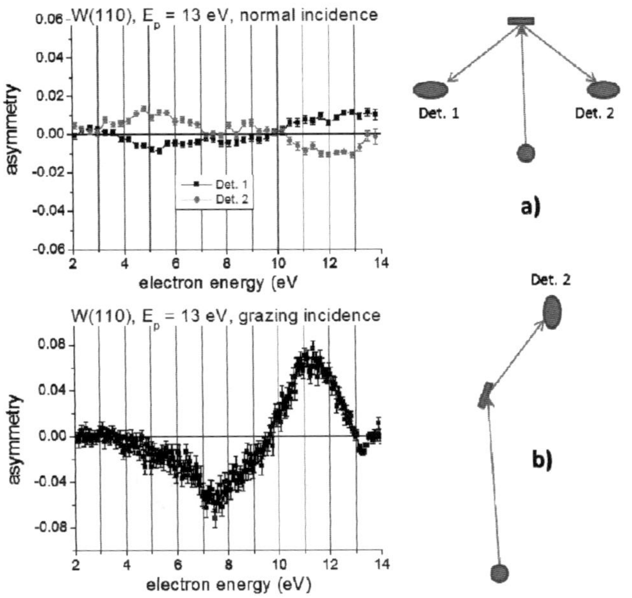

Figure 3. a) Asymmetry of the secondary emission spectrum from W(110) for E_p = 13 eV, normal incidence; b) asymmetry for E_p = 13 eV, grazing incidence ($\theta = 70^0$).

For specular geometry and grazing incidence the asymmetry of the secondary emission spectrum (including true secondaries, elastically and inelastically scattered primaries) generated by the 13 eV spin-polarized electron beam is very different from the normal incidence that is shown in figure 3 b).

The origin of the intensity asymmetry of secondary emission spectra excited by spin-polarized electrons in this case is the so called "spin-filter effect" [25, 26]. The qualitative explanation of this effect is the following. Incident spin-polarized electrons while losing their energy via electron-electron scattering still keep a certain degree of polarization. On the way out of the solid to the vacuum these electrons undergo spin-dependent (due to spin-orbit interaction) scattering at the surface, such that their transmission through the surface potential barrier becomes spin-dependent. Presumably incident electrons also undergo spin-dependent transmission through the surface penetrating into the surface region of the crystal. It is seen from figure 3 that the intensity asymmetry for grazing incidence is even larger than for normal incidence. These features in the asymmetry spectrum are related to the scattering events with large energy and momentum transfer. For the specular geometry the number of such event is relatively small since the elastic maximum is dominating in the spectrum at this geometry. On the other hand these features in the intensity asymmetry highlight the contribution of electron-electron collisions in contrast to the dipole mechanism of the electron-surface interaction.

3.2. Ferromagnetic film of Fe on W(110)

In contrast to the spin-orbit interaction the exchange effect depends on the relative orientation of the polarization vector of the incident beam and the magnetic moment of the magnetized sample. For example, figure 4 shows an asymmetry of the Stoner excitations in a 3ML Fe film on W(110) excited by normal incidence spin-polarized electrons for two opposite orientations of magnetic moment of the sample (solid circles and open circles). These two asymmetries are mirror symmetric with respect to the X- axis. In addition, the asymmetry of SPEELS was measured for the magnetic moment of the sample perpendicular to the polarization vector of the incident beam (triangles).

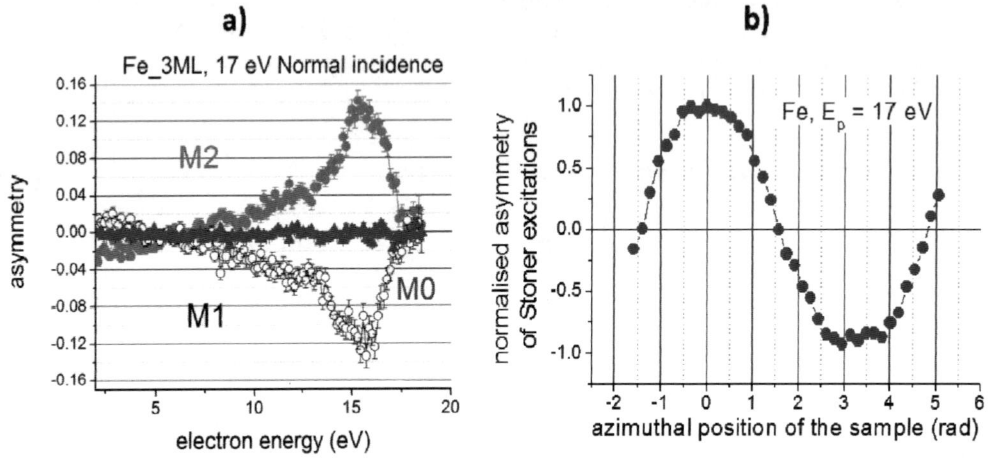

Figure 4. a) Stoner excitations asymmetry for three orientations of the magnetic moment **M** of the sample; b) normalized asymmetry of Stoner's excitations as a function of the azimuthal position of the sample, zero angle corresponds to the collinear orientation of the magnetic moment of the Fe film (**M**1) and polarization vector of the incident beam.

This asymmetry apparently is zero. Systematic measurements of the asymmetry of Stoner excitations on the Fe film as a function of the azimuthal angle of the sample (i.e. the angle between the magnetic moment of the sample and polarization vector of the incident beam) shows an almost perfect cosine dependence (figure 4b). These measurements confirm that the magnitude of the Stoner excitation asymmetry in case of a single domain ferromagnetic sample is proportional to the scalar product (**M**·**P**) of the magnetization (**M**) of the sample and the polarization vector (**P**) of the incident beam.

The origin of the asymmetry in the above spectra is the different probability of spin flip exchange process for spin-up and spin-down incident electrons. Each of these scattering processes leads to the creation of an electron-hole pair with opposite spin characters, which is called Stoner excitation. A probability of an energy loss (Stoner excitation) with a corresponding momentum transfer in SPEELS depends on the number of available combinations (pairs) of states in occupied and unoccupied bands of opposite spin character (Density of Stoner states) and a matrix element of a corresponding transitions. The density of Stoner states is given by [27]:

$$\rho_{nn'}(E) = \frac{1}{N_0}\sum_{\mathbf{k}}(f_{n'\mathbf{k}-\mathbf{q}\sigma_f} - f_{n\mathbf{k}\sigma_i})\cdot\delta(E + \varepsilon_{n'\sigma_f}(\mathbf{k}-\mathbf{q}) - \varepsilon_{n\sigma_i}(\mathbf{k})) \quad , \qquad (1)$$

where n and n' are the indices of majority and minority bands, $\mathbf{k}$ is the Bloch wave vector, $f_{n\mathbf{k}\sigma}$ is the occupation probability of the state specified by $(n,\mathbf{k},\sigma)$, $\mathbf{q}$ is the momentum transfer. The δ-function ensure the energy and momentum balance in the density of Stoner states.

The matrix element of the interband transition (between a majority and a minority bands) in given by [27]:

$$W_{nn'}(\mathbf{p}_i,\mathbf{q},\mathbf{k}) = \frac{N_0}{V}\sum_{\mathbf{K}} \hat{\text{v}}(\mathbf{k}-\mathbf{p}_i-\mathbf{K})\cdot\hat{\phi}^*_{n\mathbf{k}}(\mathbf{k}-\mathbf{K})\cdot\hat{\phi}_{n'\mathbf{k}-\mathbf{q}}(\mathbf{k}-\mathbf{q}-\mathbf{K}) \ , \qquad (2)$$

where $\mathbf{p}_i$, $\mathbf{q}$ and $\mathbf{k}$ are the incident electron momentum, momentum transfer and Bloch wave vector of the valence state, respectively; **K** is the reciprocal lattice vector; V is the volume of the crystal; N_0 is the number of sites; functions v and ϕ with caret signs are the Fourier transforms of the interaction potential and the Wannier functions, respectively. As was pointed out in [28], for a fixed scattering angle, i.e., for given incoming and outgoing momenta, there are three possible scattering processes corresponding to different momentum transfer: i) in the first one a specular scattering follows a small momentum transfer inelastic scattering giving momentum transfer **q**; ii) in the second, an elastic (or specular) scattering precedes a relatively small-angle inelastic scattering event, giving momentum transfers **q'**; iii) in the third process, there is a single large-angle inelastic scattering event with a momentum transfer **q"** absorbed by the electron-hole pair excitations. These types of scattering may occur in all three geometrical arrangements shown in figure 2. The processes i) and ii) are explicitly shown only in figure 2a) and c), respectively. Which process with which momentum transfer **q**, **q'** or **q"** contributes mostly to the observed SPEELS of Stoner excitations depends on the kinematics of the scattering. The partial cross sections of Stoner excitations corresponding to these three possible scenarios of momentum transfer were calculated in [28]. It was found for a specular geometry that the cross section corresponding to **q"** momentum transfer (direct electron-electron encounter) makes a dominant contribution to the total cross section. At off-specular geometry the cross sections corresponding to **q'** and **q"** momentum transfer are almost equal and much larger than the cross section corresponding to **q** momentum transfer. The overall shape of Stoner excitations in SPEELS is determined by the combined effect of the density of Stoner states (1) and the matrix element (2).

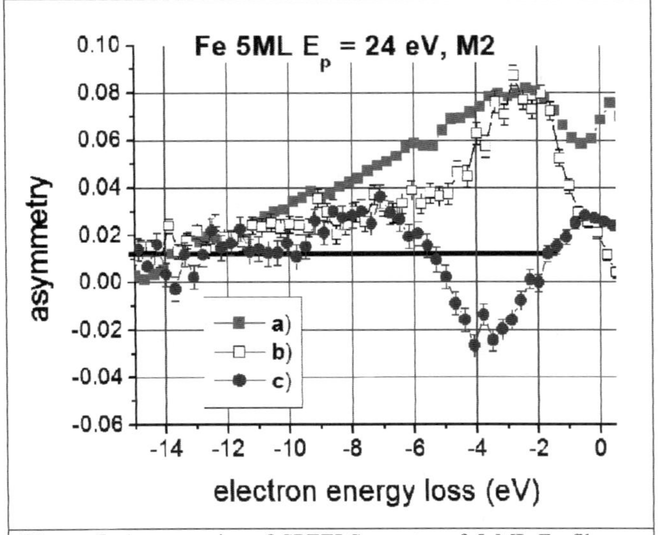

Figure 5. Asymmetries of SPEELS spectra of 5 ML Fe film on W(110) measured at various geometries depicted on figure 1 as a), b), and c).

Figure 5 represents SPEELS asymmetry spectra of 5 ML iron film on W(110) excited by 24 eV spin-polarized electrons at three different kinematics: a) specular geometry with the incidence angle of 25^0: b) normal incidence and 50^0 detection angle; c) specular geometry with 70^0 incident angle. First two spectra show a positive asymmetry with a maximum at about 2.5 eV of energy loss. This is consistent with the Stoner excitation maximum position. The sign of asymmetry corresponds to the magnetization M2 in our nomenclature. The shapes of these two asymmetry spectra are quite different: for specular geometry the asymmetry of elastic scattering has almost the same value as the Stoner excitation asymmetry and there are two kinks at about 5 eV and 9 eV of energy loss. For normal incidence the asymmetry of elastic scattering is almost zero and the Stoner excitation asymmetry is well localised at 2.5 eV. A very different asymmetry is observed for the 70^0 of incidence: first of all the sign of the asymmetry is changing from negative, around the Stoner excitation energy loss, to a positive sign at larger energy losses. We note here that the magnetization of the film is the same (M2). It appears that there is an overlap (sum) of two contributions to the asymmetry spectrum at the grazing incidence: the Stoner excitation asymmetry with the positive sign, similar to geometries a) and b), and a negative maximum of asymmetry located at about 4 eV energy losses, which is due to a different from Stoner excitation mechanism. Our preliminary investigation shows that this maximum in the asymmetry spectrum is related rather to an emission feature in the secondary electron emission spectrum. The origin of this maximum needs further investigation.

3.3. Spin-polarized two-electron spectra of Co film on W(110)

In the two-electron spectroscopy in reflection geometry, two low-energy electrons resulting from the

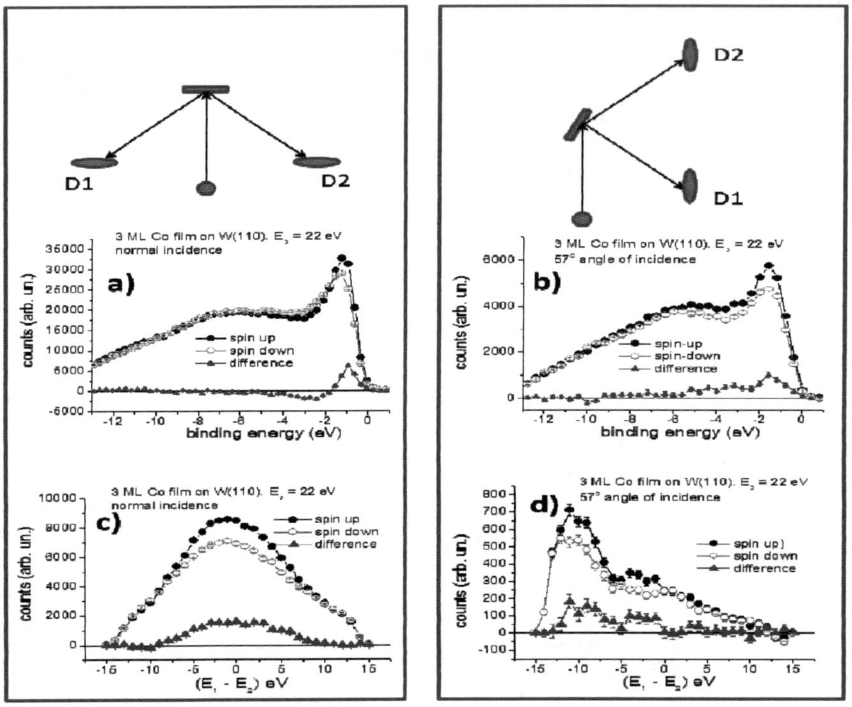

Figure 6. Binding energy spectra a) and b) and energy sharing distributions c) and d) of 3 ML cobalt film on W(110). Primary electron energy 2 4 eV. Normal incidence – left panel; 57^0 angle of incidence – right panel.

interaction of a single incident electron with a surface are detected in coincidence and their energy and momenta are measured. The energy distribution of correlated electron pairs is called the (e,2e) spectrum. Using a spin-polarized incident beam the (e,2e) spectra are measured for spin-up and spin-down incident electrons. Figure 6 illustrates the influence of the geometry of the experiment on the shape of (e,2e) energy distributions and on spin effect in the (e,2e) reaction. The left panel of figure 6 shows binding energy spectra and energy sharing distributions recorded at normal incidence of 24 eV spin-polarized electron beam for 3 ML cobalt film on W(110). The binding energy according to the energy conservation in the (e,2e) reaction is determined by the relation: $E_b = E_0 - E_1 - E_2$, where E_0, E_1, E_2 are the energies of incident and two scattered electrons, respectively. The angle between detectors is 100^0. Spectra measured for spin up and spin down polarization of the incident beam are shown together with the difference spectrum (triangles in the figure). The right panel shows binding energy spectra and energy sharing distributions measured on 3 ML cobalt film on W(110) at 57^0 angle of incidence. The angle between detectors is 90^0 and the angle between incident beam and detector D1 is 50^0. The shapes of binding energy spectra for both geometrical arrangements are similar (see figure 6 a) and b)) whereas the energy sharing distributions (figure 6 c) and d)) are very different. The difference in the energy sharing distributions for normal and 57^0 of incidence is explained by the fact that the tangential component of the electron momentum is conserved. Indeed, for the off-normal incidence the total momentum of the pair of electrons is carrying parallel to the surface momentum of the incident electron: $\mathbf{K}_{b||} + \mathbf{K}_{0||} = \mathbf{K}_{1||} + \mathbf{K}_{2||}$, where $\mathbf{K}_b$, $\mathbf{K}_0$, $\mathbf{K}_1$, $\mathbf{K}_2$ are momentum of the bound (valence) electron, incident electron and two outgoing electrons, respectively. The spin effects (difference spectra), for both binding energy spectra and energy sharing distributions, at two geometries are different. The asymmetry of the binding energy spectrum to a large extent depends on the spin asymmetry of the Bloch spectral density function of the sample [29]. It is evident from figure a) and b) that for the off-normal incidence the asymmetry of the binding energy spectrum (even the sign) depends also on the scattering dynamics. The difference spectrum of the energy sharing distribution for the normal incidence and symmetric detection of pairs shows a maximum in the middle of the spectrum (energies of electrons are almost equal), whereas for off-normal incidence the difference spectrum has complicated structure and asymmetric with respect to the zero point ($E_1 - E_2 = 0$).

4. Conclusions

The experimental results indicate that spin effects in inelastic (and elastic) scattering of spin-polarized electrons from magnetic and nonmagnetic surfaces depend very much on the geometrical arrangement of the experiment, i.e. on angle of incidence and angle of detection (in SPEELS), as well as on angle between detectors (in spin-polarized two-electron spectroscopy). We note that for grazing incidence of spin polarized electrons on 5 ML Fe film on W(110) a new mechanism responsible for intensity asymmetry in the secondary emission spectrum is confirmed. In (e,2e) the intensity asymmetry in binding energy spectra and in energy sharing distributions is determined not only by an imbalance of spin-up and spin-down states in the valence band but also by kinematics.

Acknowledgments

This research was supported by the Australian Research Council and the University of Western Australia. OMA acknowledges Saint-Petersburg State University for a research grant 11.38.187.2014. We thank S. Key and G. Light (UWA) for their technical support.

References

[1] Kirschner J 1985 *Phys. Rev. Lett.* **55** 973
[2] Hopster H J 1985 *Appl. Phys.* **57** 8, pt.2A, 3017
[3] Kirschner J and Suga S 1986 *Surf. Sci.* **178** 327-335
[4] Modesti S, Dellavalle F, Rosei R *et al.* 1985 *Phys. Rev. B* **31** 5471
[5] Kirschner J 1985 *Surf. Sci.* **162** 83
[6] Hopster H, Abraham D L and Pappas D P 1990 *J. Electr. Spectr. and Rel. Phen.* **51** 301
[7] Venus D and Kirschner J 1988 *Phys. Rev. B* **37** 2199
[8] Kirschner J and Suga S 1987 *Solid State Communications* **64** 997
[9] Vollmer R, Etzkorn M, Anil Kumar P S, Ibach H and Kirschner J 2003 *Phys. Rev. Lett.* **91** 147201-1
[10] Prokop J, Tang W X, Zhang Y, Tudosa I, Peixoto T R F, Zakeri Kh and Kirschner J 2009 *Phys. Rev. Lett.* **102** 177206
[11] Samarin S, Artamonov O M, Sergeant A D and Williams J F 2005 *Phys. Rev. B* **72** 235419
[12] Samarin S, Artamonov O M, Sergeant A D, Stamps R and Williams J F 2006 *Phys. Rev. Lett.* **97** 096402
[13] Pierce D T, Meier F and Zürcher P 1980 *Rev. Sci. Instrum.* **51** 478
[14] Samarin S N, Artamonov O M, Waterhouse D K, Kirschner J, Morozov A and Williams J F

2003 *Rev. Sci. Instr.* **74** 1274
[15] Samarin S N, Artamonov O M, Sergeant A D, Pravica L, Cvejanovic D, Wilkie P, Guagliardo P, Suvorova A A and Williams J F 2009 *J. Phys.: Conf. Ser.* **185** 012041
[16] Samarin S N, Artamonov O M, Sergeant A D and Williams J F 2007 *Surf. Sci.* **601** 4343
[17] Samarin S, Berakdar J, Artamonov O M and Kirschner J 2000 *Phys. Rev. Lett.* **85** 1746
[18] Zakeri Kh, Peixoto T R F, Zhang Y, Prokop J and Kirschner J 2010 *Surf. Sci.* **604** L1
[19] Cortenraada R, Ermolov S N , Semenov V N, Denier van der Gon A W, Glebovsky V G, Bozhko S I and Brongersma H H 2001 *J. Cryst. Growth* **222** 154
[20] Willis Roy F and Christensen N E 1978 *Phys. Rev. B* **18** 5140
[21] Christensen N E and Willis Roy F 1979 *J. Phys. C: Solid State Phys.* **12** 167
[22] Zürcher P, Meier F and Christensen N E 1979 *Phys. Rev. Lett.* **43** 54
[23] Samarin S, Artamonov O M, Sergeant A D and Williams J F 2005 *Surf. Sci.* **579** 166
[24] Samarin S, Artamonov O M, Sergeant A D, Kirschner J, Morozov A and Williams J F 2004 *Phys. Rev. B* **70** 073403
[25] Oepen H P, Hünlich K and Kirshner J 1986 *Phys. Rev. Lett.* **56** 496
[26] Garbe J and Kirschner J 1989 *Phys. Rev. B* **39** 9859
[27] Saniza R and Apell S P 2000 *Phys. Rev. B* **63** 014409
[28] Saniza R and Apell S P 1993 *Phys. Rev. B* **48** 3206
[29] Morozov A, Berakdar J, Samarin S N, Hillebrecht F U and Kirschner J 2002 *Phys. Rev. B* **65** 104425

Two-color above-threshold and two-photon sequential double ionization beyond the dipole approximation

A N Grum-Grzhimailo[1], E V Gryzlova[1], E I Kuzmina[1,2], A S Chetverkina[1,2] and S I Strakhova[1]

[1]Skobeltsyn Institute of Nuclear Physics, Lomonosov Moscow State University, Moscow 119991, Russia

[2]Faculty of Physics, Lomonosov Moscow State University, Moscow 119991, Russia

E-mail: algrgr1492@yahoo.com

Abstract. Two nonlinear atomic photoprocesses are theoretically considered with the emphasis on the photoelectron angular distributions and their modifications due to violation of the dipole approximation: sequential two-photon double ionization and two-color above threshold ionization. These reactions are now accessible with X-ray free electron lasers. Both processes are exemplified by the ionization of krypton: from the 4p shell in the sequential two-photon double ionization and from the 2s shell in the two-color above-threshold ionization, which are compared to the Ar(3p) and Ne(1s) ionization, respectively. Noticeable nondipole effects are predicted.

1. Introduction

Recent advances in development and use of free-electron lasers (FELs) producing very intense X-ray and VUV radiation made it possible to observe angular distributions of emitted electrons in non-linear atomic photoprocesses in this energy range (for example, [1, 2, 3, 4]). It is known that the photoelectron angular distributions (PADs) are sensitive to the interference between photoionization amplitudes of different multiplicity and parity and that this interference (dominantly between the electric dipole E1 and electric quadrupole E2 amplitudes) can influence the PAD in the one-photon single ionization already in the VUV region leading to pronounced non-dipole effects [5, 6, 7]. Naturally, the non-dipole effects should show up for non-linear photoprocesses in a similar energy range. Thus, motivated by the achievements in using the X-ray FELs, we have developed a theory for the PADs and angular correlations between the two emitted electrons in sequential two-photon double ionization (2PDI) [8] (figure 1a) and the PADs for two-color above-threshold ionization (ATI) [9] of atoms (figure 1b) beyond the dipole approximation. In the latter case, the XUV photon generates the nondipole transitions to the atomic continuum which are then monitored by the optical (IR) radiation.

Numerical predictions for the non-dipole effects were made for the 2PDI of the outer np^6 shells of neon and argon by linearly polarized radiation [8, 10] and of neon, argon and krypton for circularly polarized radiation [11]. In section 2 of this contribution we 'fill the gap' and present the results for the 2PDI of the $4p^6$ shell in krypton by linearly polarized FEL. An extra Cooper minimum in the quadrupole cross section leads to differences in the general behavior of PADs in sequential 2PDI of Kr and Ar from 4p and 3p shells, respectively. In section 3 we give

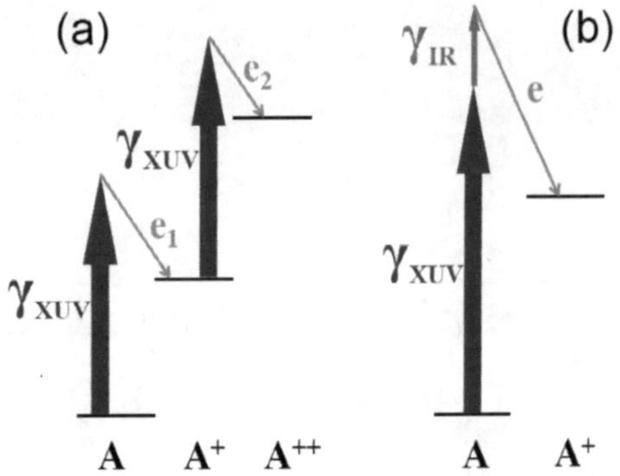

Figure 1. Schemes of sequential 2PDI (a) and two-color ATI (b).

predictions for the two-color XUV + optical laser ATI of the 2s shell in atomic Kr and compare the results with the similar process for the 1s ionization in atomic Ne.

We assume that the photon pulses contain many optical cycles and that the photon beams are either mutually incoherent (in the two-color ATI) or the coherence is lost between the two steps of the process (2PDI). Under these conditions, the effects of the pulse shape, relative phase of the electromagnetic fields and other effects for which the time dependence of the electromagnetic field is crucial, are irrelevant in treatment of the PADs. Furthermore, in the lowest order pertubation theory, the PADs are independent of the beam intensities.

2. Sequential two-photon double ionization

The PADs in the sequential 2PDI were studied, both experimentally and theoretically, mostly for the outer shells of the noble gases. In a stepwise approximation this reaction proceeds in two steps,

$$\gamma \; + \; \mathrm{A(np^6)} \longrightarrow \mathrm{A^+(np^{5\,2}P_{1/2,3/2})} \; + \; e_1\,; \tag{1}$$

$$\gamma \; + \; \mathrm{A^+(np^{5\,2}P_{1/2,3/2})} \longrightarrow \mathrm{A^{++}(np^{4\,3}P,\, ^1D,\, ^1S)} \; + \; e_2\,, \tag{2}$$

with possible evolution of the polarized intermediate ion $\mathrm{A^+(np^{5\,2}P_{1/2,3/2})}$ between the two steps. Electrons e_1 and e_2 are well distinguished by their kinetic energy. Expressions for the angular distributions and correlations of the emitted electrons e_1 and e_2 in the sequential 2PDI have been obtained in [8] within the statistical tensor formalism taking into account the full multipole expansion of radiation in electric and magnetic moments. The formalism was further specified for linearly polarized [8] and circularly polarized [11] FEL beams with inclusion of the first-order corrections to the dipole approximation due to the E1-E2 interference of the multipole amplitudes.

We concentrate here on the PAD of the second electron e_2. For a linearly polarized FEL beam it is of the form

$$\begin{aligned} \frac{d\sigma}{d\Omega} \;=\; & \frac{\sigma}{4\pi}\Big(1 + \beta_2^{(2)}\,P_2(\cos\vartheta) + \beta_4^{(2)}\,P_4(\cos\vartheta) \\ & + \Big(\delta^{(2)} + \gamma_2^{(2)}\cos^2\vartheta + \gamma_4^{(2)}\cos^4\vartheta\Big)\sin\vartheta\cos\varphi\Big)\,, \end{aligned} \tag{3}$$

where σ is the angle integrated ionization cross section in the dipole approximation (the E1-E2 interference terms vanish in the angle integrated cross section), the z-axis of the coordinate system is directed along the electric field (polarization vector) of the FEL beam, which propagates along the x-axis. The first three terms and the asymmetry parameters $\beta_2^{(2)}$, $\beta_4^{(2)}$

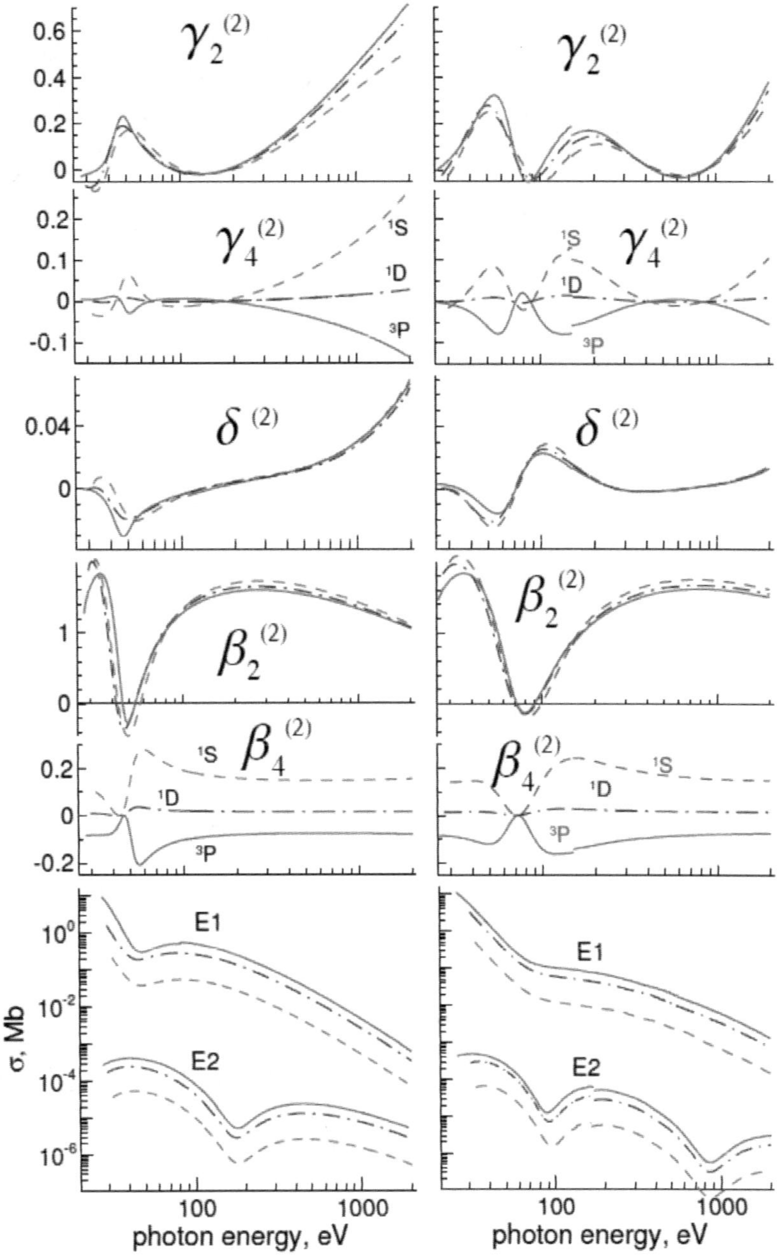

Figure 2. Dipole (E1) and quadrupole (E2) cross sections (bottom panels) and asymmetry parameters of the PAD (3) emitted in sequential 2PDI of the valence $3p^6$ shell of Ar (left column) and $4p^6$ shell of Kr (right column) for different *LS* terms of the residual doubly charged ion $A^{++}(np^4)$ as a function of the photon energy: ^3P (solid red), ^1D (chain blue), ^1S (dashed green). See text for an explanation of the discontinuity in some curves for Kr at about 150 eV.

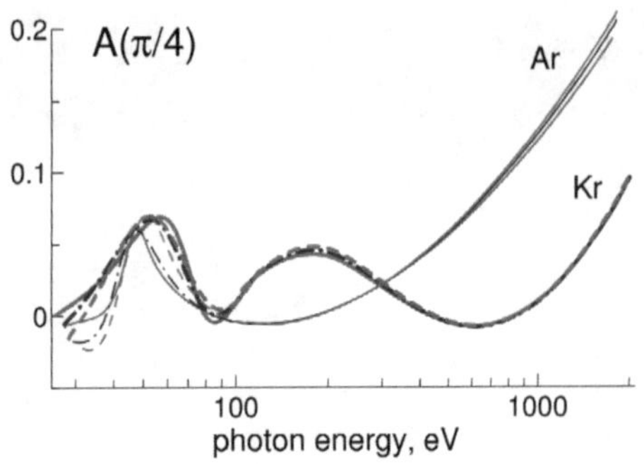

Figure 3. Differential forward-backward asymmetry (4) for $\vartheta = \frac{\pi}{4}$ as a function of the photon energy for different final multiplet states of $Kr^{++}(4p^4)$ (bold lines) and $Ar^{++}(3p^4)$ (thin lines) at 2PDI. Curves as in figure 2.

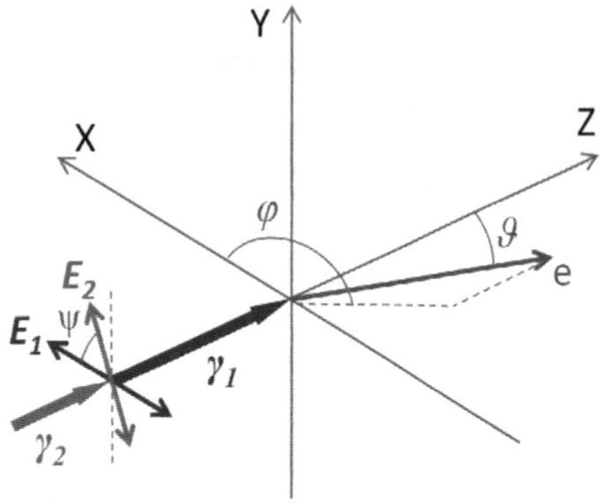

Figure 4. Geometry of the two-color ATI by two parallel linearly polarized radiation beams γ_1 and γ_2. Vectors $\mathbf{E_1}$ and $\mathbf{E_2}$ denote the corresponding directions of the linear polarization.

correspond to the pure dipole contribution; the last three terms with the parameters $\delta^{(2)}$, $\gamma_2^{(2)}$, $\gamma_4^{(2)}$ give the first-order non-dipole corrections and appear due to the E1-E2 interference.

The asymmetry parameters in equation (3) are expressed in terms of the photoionization amplitudes [8]. The atomic model used in the calculations of the matrix elements is based on the multiconfiguration Hartree-Fock (MCHF) approach and has been described for Kr in [12]. Excitation of the intermediate Kr^+ $4p^5\,{}^2P_{1/2,3/2}$ fine-structure ionic states is treated as incoherent and the corresponding contributions to the cross sections and PADs are summed up, as well as the contributions from unresolved fine-structure states of the final doubly charged ion Kr^{++} $4p^4\,{}^3P_{0,1,2}$. The intermediate ionic spin-orbit doublets of $Ar^+(3p^5)$, and $Kr^+(4p^5)$ are splitted by 0.177 eV, and 0.656 eV, respectively. As it was confirmed by calculations of Nikolopoulos [13], the time parameter defining the coherency of the excitation is the spin-orbit precession time: 23.30 fs for Ar^+, and 6.21 fs for Kr^+. Thus for Kr, the excitation can be treated as incoherent for almost all pulses currently generated by FELs in the XUV.

The five upper double panels of figure 2 show the asymmetry parameters in equation (3) for the sequential 2PDI of the argon 3p and krypton 4p shell. The bottom panels display the corresponding angle integrated dipole (E1) and quadrupole (E2) ionization cross sections. The results for Ar [10] are shown for comparison and the scales of the corresponding plots for Ar and Kr are identical. For high enough photon energies, the difference between the MCHF and single-configuration calculations becomes negligible. For these energies we used therefore the single-configuration model, which provides better stability in calculations of the continuum wavefunctions. This explains the tiny jumps in some curves for Kr in figure 2 at about 150 eV.

The pure dipole parameters $\beta_2^{(2)}$ and $\beta_4^{(2)}$ for Ar and Kr qualitatively behave similar as functions of the photon energy. The Cooper minimum in the dipole np→Ed amplitude is the

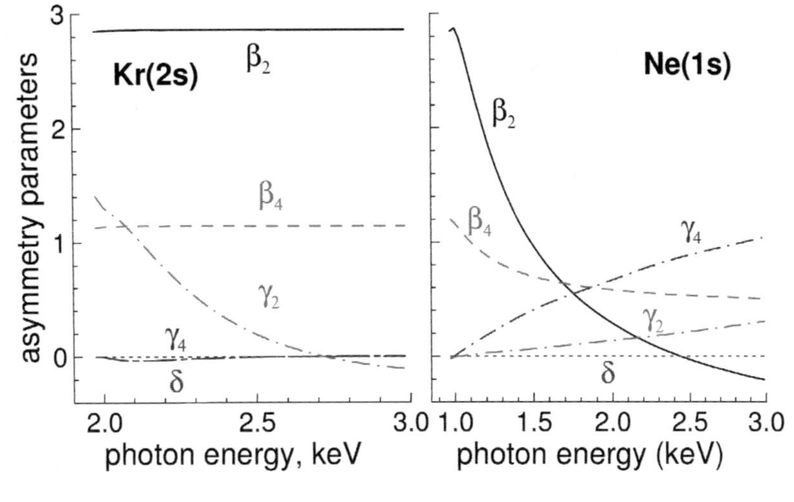

Figure 5. Angular asymmetry parameters of equation (3) for the two-color Kr(2s) (left) and Ne(1s) (right) ATI by linearly polarized in the same direction photons.

reason for sharp changes in the parameters. This minimum, seen in the E1 curves for the cross sections, is filled due to the np→Es ionization transitions. As has been discussed in detailed in [10] for the case of Ar, the decrease of the dipole ionization causes an increase of the non-dipole asymmetry parameters. In contrary, the Cooper minima in the quadrupole np→Ef amplitude, located around 200 eV photon energy for Ar^+ and around 100 eV and 800 eV for Kr^+, lead to negligible values of $\delta^{(2)}$, $\gamma_2^{(2)}$, $\gamma_4^{(2)}$ in these regions. The two quadrupole Cooper minima in Kr^+ against the single one in Ar^+ cause differences in the energy dependence of the nondipole asymmetry parameters for these two atoms: while in Ar the absolute values of the parameters increase monotonically with photon energy starting from 200 eV, the parameters in Kr show a second minimum at higher energies.

Similar features are demonstrated in figure 3 for the differential forward-backward asymmetry (in the xz plane)

$$A(\vartheta) = \left[\frac{d\sigma}{d\Omega}(\vartheta) - \frac{d\sigma}{d\Omega}(\pi - \vartheta)\right] \Big/ \left[\frac{d\sigma}{d\Omega}(\vartheta) + \frac{d\sigma}{d\Omega}(\pi - \vartheta)\right] \tag{4}$$

for $\vartheta = \frac{\pi}{4}$. The forward-backward asymmetry is a convenient quantity to study nondipole effects, since it vanishes in the dipole approximation. Noteworthy, the asymmetries (4) for sequential 2PDI in both Ar and Kr are similar for $np^4\ {}^3P$, ${}^1D$, and ${}^1S$ final states, except near the threshold. Therefore it is experimentally not required to resolve these multiplet states by the electron spectrometer for high electron energies, where the nondipole effects are generally larger. At the same time, an asymmetry of more than 5% is predicted for both Ar and Kr at low photon energies of 50-60 eV in the region of the Cooper minimum for the np→Ed transition.

The non-dipole effects notably change the PADs. In the Cooper minimum for the np ionization of Ar^+ and Kr^+ at photon energies of tens of eV, the predicted non-dipole effects are as pronounced as at photon energies of about 1 keV, but the conditions of their observation are better in the Cooper minimum due to the much larger photoionization cross sections at lower photon energies [10].

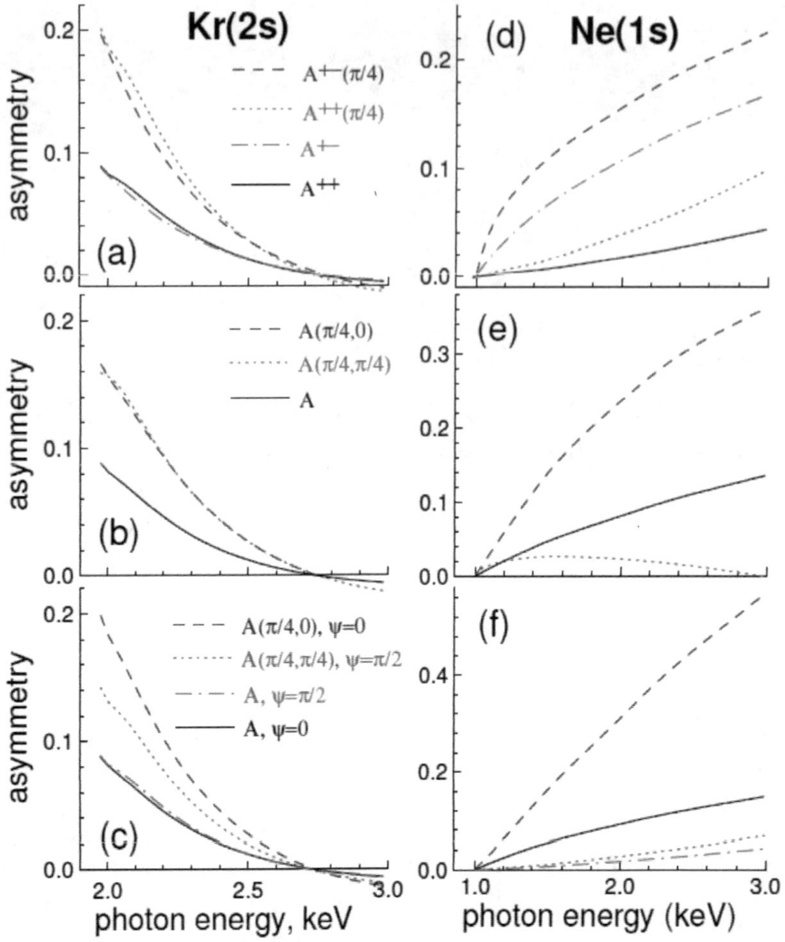

Figure 6. Forward-backward differential asymmetries $A(\vartheta, \varphi)$ (4), and integral asymmetries A (5), for circularly polarized XUV and IR photons (a) and (d), linearly polarized XUV photon and circularly polarized IR photon (b) and (e), and linearly polarized XUV and IR photons (c) and (f). Left column: for Kr(2s) ATI; right column: for Ne(1s) ATI.

3. Two-color above-threshold ionization

Here we apply the formalism for the PADs in two-color ATI to the experimentally most feasible arrangement with collinear photon beams and emphasize independently variable polarizations of the radiation beams [9]. The geometry for two linearly polarized beams is shown in figure 4. Particular expressions for the PADs within the first-order non-dipole corrections have been derived in [9]. In particular, for parallel polarizations ($\psi = 0$), the PAD in the ATI line is described by an equation similar to (3). To avoid confusion, for the ATI (i.e. single ionization) we will use another notations for the asymmetry parameters in (3): $\beta_2^{(2)} \to \beta_2$, $\beta_4^{(2)} \to \beta_4$, $\delta^{(2)} \to \delta$, $\gamma_2^{(2)} \to \gamma_2$, $\gamma_4^{(2)} \to \gamma_4$.

Numerical results for the amplitudes of the two-color Kr(2s) ATI were obtained here within the second-order perturbation theory similar to the treatment of Ne(1s) in [9]. The energy of the optical photon was fixed at 1.5 eV. The ionization thresholds of the 2s shell in Kr and of the 1s shell in Ne are 1923 eV and 870 eV, respectively. It is known from the calculations [6] that the nondipole effects in the PAD for the single-photon Kr(2s) ionization are large close to the ionization threshold and with increasing photon energy they are first decreasing. The latter feature is not usual, since normally the relative contribution of the higher multipoles of the radiation increases with increasing photon energy leading also to the increasing non-dipole effects (like in the Ne(1s) ionization [6]). Figure 5 shows that similar counter trends persist in the two-color Kr(2s) and Ne(1s) ATI: the nondipole parameter γ_2 is decreasing for Kr(2s), while the nondipole parameters γ_2 and γ_4 are increasing for Ne(1s). The calculations reveal another interesting feature of the dipole parameters β_2 and β_4: they strongly vary with photon energy for the Ne(1s) ATI and are nearly constants for the Kr(2s) ATI. Our analysis shows that the nearly constant values of the parameters β_2 and β_4 follow from the dynamical calculations for the Kr(2s) ATI and are related to an 'accidentally' almost constant absolute ratio of the Es and Ed two-photon ionization amplitudes in the broad range of photon energies. This feature is supposed to be model-dependent.

The differential forward-backward asymmetry (4) for the case of one circular and one linear polarized beam depends on the azimuthal angle φ, $A = A(\vartheta, \varphi)$. For two linearly polarized beams, the asymmetry depends additionally on the adjustable angle ψ between the polarization directions of the photon beams (see figure 4). The integral forward-backward asymmetry is defined as

$$A = \frac{\sigma_{\Omega+} - \sigma_{\Omega-}}{\sigma_{\Omega+} + \sigma_{\Omega-}}, \tag{5}$$

where $\sigma_{\Omega+}$ ($\sigma_{\Omega-}$) is the photoelectron flux integrated over the forward (backward) hemisphere. Figure 6 shows various types of forward-backward asymmetry as function of the photon energy. Comparing the Kr(2s) and Ne(1s) ATI, it is clearly seen that, again, all asymmetries for Kr(2s) are maximal at the ionization threshold and drop down with increasing photon energy, while the asymmetries for Ne(1s) are negligible at the ionization threshold and increase with increasing photon energy. At the photon energies considered here, the asymmetries for Kr(2s) are not larger than 0.2, which is smaller than some of the asymmetries for Ne(1s) (figure 6).

4. Conclusion

We predicted noticeable nondipole effects in two-photon two-color above-threshold Kr(2s) ionization and in sequential two-photon Kr(4p) double ionization. Some new features of the processes are predicted in comparison with the Ne(1s) and Ar(3p) ionization, respectively. The non-dipole effects can be observed with existing FEL facilities and combining XUV FEL and optical laser beams. We hope that the results presented will stimulate experiments on non-dipole effects in nonlinear atomic photoprocesses.

Acknowledgments

The authors acknowledge support by the Russian Foundation for Basic Research (RFBR) under Grant No. 12-02-01123. EVG gratefully acknowledges financial support by the Dynasty Foundation via the Support Program for Young Scientists. Fruitful discussions with Dr Michael Meyer are gratefully acknowledged.

References

[1] Berrah N *et al* 2010 *J. Mod. Opt.* **57** 1015
[2] Bostedt C *et al* 2013 *J. Phys. B: At. Mol. Opt. Phys.* **46** 164003
[3] Feldhaus J *et al* 2013 *J. Phys. B: At. Mol. Opt. Phys.* **46** 164002
[4] Yabashi M *et al* 2013 *J. Phys. B: At. Mol. Opt. Phys.* **46** 164001
[5] Guillemin R, Hemmers O, Lindle D W and Manson S T 2006 *Radiat. Phys. Chem.* **75** 2258
[6] Trzhaskovskaya M B, Nefedov V I and Yarzhemsky V G 2001 *Atomic Data and Nuclear Data Tables* **77** 97

[7] Kleinpoppen H, Lohmann B and Grum-Grzhimailo A N 2013 *Perfect/Complete Scattering Experiments* (Berlin: Springer) p 283
[8] Grum-Grzhimailo A N, Gryzlova E V and Meyer M 2012 *J. Phys. B: At. Mol. Opt. Phys.* **45** 215602
[9] Grum-Grzhimailo A N and Gryzlova E V 2014 *Phys. Rev. A* **89** 043424
[10] Gryzlova E V, Grum-Grzhimailo A N, Strakhova S I and Meyer M 2013 *J. Phys. B: At. Mol. Opt. Phys.* **46** 164014
[11] Gryzlova E V, Grum-Grzhimailo A N, Kuzmina E I and Strakhova S I 2014 *J. Phys. B: At. Mol. Opt. Phys.* **47** 195601
[12] Fritzsche S, Grum-Grzhimailo A N, Gryzlova E V and Kabachnik N M 2009 *J. Phys. B: At. Mol. Opt. Phys.* **42** 145602
[13] Nikolopoulos L A A 2013 *Phys. Rev. Lett.* **111** 093001

AUTHOR INDEX